REMOTE SENSING

IN ECOLOGY

Remote Sensing
in Ecology

Edited by
PHILIP L. JOHNSON

UNIVERSITY OF GEORGIA PRESS

ATHENS

Contents

Contributors

Dr. Robert C. Aldrich
Pacific Southwest Forest and
Range Expt. Sta.
P. O. Box 245
Berkeley, California 94701

Dr. Reid A. Bryson
Department of Meteorology
University of Wisconsin
Madison, Wisconsin 53706

Mr. William E. Butler
School of Natural Resources
University of Michigan
Ann Arbor, Michigan 48104

Dr. George L. Clarke
Biological Laboratories
Harvard University
Cambridge, Massachusetts 02138

Mr. G. Ross Cochrane
Department of Geography
University of Kansas
Lawrence, Kansas 66044

Dr. Alfred Conrod
Building N-51
Mass. Inst. of Technology
Cambridge, Massachusetts 02139

Dr. Charles F. Cooper
School of Natural Resources
University of Michigan
Ann Arbor, Michigan 48104

Dr. David M. Gates
Missouri Botanical Garden
2315 Tower Grove Avenue
St. Louis, Missouri 63110

Dr. Robert C. Heller
Pacific Southwest Forest and
Range Expt. Sta.
P. O. Box 245
Berkeley, California 94701

Dr. Roger M. Hoffer
Laboratory for Agricultural
Remote Sensing
Purdue University
West Lafayette, Indiana 47906

Mr. Christian V. Johannsen
Laboratory for Agricultural
Remote Sensing
Purdue University
West Lafayette, Indiana 47906

Dr. Mahlon G. Kelly
Department of Biology
New York University
University Heights, N. Y. 10453

Dr. Edward B. Knipling
U. S. Department of Agriculture
Soil and Water Conservation
Research Div.
P. O. Box 14503,
University Station
Gainesville, Florida 32601

Mr. William A. Malida
Willow Run Laboratories
University of Michigan
Ann Arbor, Michigan 48104

Dr. R. Larry Marchinton
School of Forest Resources
University of Georgia
Athens, Georgia 30601

Dr. Dale R. McCullough
School of Natural Resources
University of Michigan
Ann Arbor, Michigan 48104

Dr. Lee D. Miller
College of Forestry and
Natural Resources
Colorado State University
Fort Collins, Colorado 80521

Dr. Stanley A. Morgan
Department of Geography
University of Kansas
Lawrence, Kansas 66044

Dr. Charles E. Olson, Jr.
School of Natural Resources
University of Michigan
Ann Arbor, Michigan 48104

Dr. Rex M. Peterson
Department of Geography
University of Kansas
Lawrence, Kansas 66044

Dr. Fabian C. Polcyn
Institute of Science and
Technology
University of Michigan
Ann Arbor, Michigan 48107

Mr. Lealand Queal
Kansas Forestry, Fish and
Game Commission
Garnett, Kansas 66032

Dr. Dave S. Simonett
Geography Department
University of Kansas
Lawrence, Kansas 66044

Mrs. Norma Spansail
Willow Run Laboratories
University of Michigan
Ann Arbor, Michigan 48104

Mr. David K. Weaver
School of Natural Resources
University of Michigan
Ann Arbor, Michigan 48104

Sondra Wenderoth
Science Engineering Research
Group
Long Island University
Greenvale, Long Island,
New York 11548

Dr. Wayne M. Wendland
Center for Climatic Research
University of Wisconsin
733 University Avenue
Madison, Wisconsin 53706

Dr. Edward Yost
Science Engineering Research
Group
Long Island University
Greenvale, Long Island,
New York 11548

Preface

THREE new technologies have appeared on the ecological scene in recent years; computers, radioisotobes and remote sensors. The first two have matured and yielded substantial contributions in many disciplines including ecology. Hardware, buildings and business enterprises have evolved to focus computer and radiation technology on a broad spectrum of problems. Remote sensing by comparison is still in its research infancy, but perhaps equal in initial cost, complexity and potential. As with any new tool or instrument we must understand how to best use it and to appreciate its limitations.

In 1967 five excellent review papers were presented to the first national symposium on remote sensing in ecology as part of the annual program of the American Institute of Biological Sciences. These review articles were summarized in the May, June, and July 1967 issues of BioScience. The possibilities of obtaining unique ecological information from airborne or orbiting platforms continues to be an exciting prospect. These reviews and previous literature have emphasized the potential of remote sensing technology for application in many landscape disciplines. An expectant attitude has been created in the scientific community. This volume represents the proceedings of a symposium held 19 June 1968 at Madison, Wisconsin, sponsored by the Ecological Society of America and the American Society of Limnology and Oceanography. The contributions in this, the second symposium, were specifically intended to emphasize research *using* remote sensing of any configuration to solve ecological problems, as opposed to conjectured future possibilities.

The diversity of these fourteen contributions demonstrates the wide applicability of this technology. The first five chapters explore the value of visible light for studying terrestrial plant communities, particularly the virtues of color, the physical basis for color differences, and some sophisticated means for analyzing color images.

Perception of emitted thermal patterns begins with a consideration
of individual plant and animal surface temperatures (Chapter 6). The
response of plant communities to a thermal gradient produced by geothermal events is treated in detail using thermal imagery (Chapter 7). Influence of time of day on thermal emissions from vegetation for optimal
imagery is briefly analyzed (Chapter 8) and the status of large vertebrate census by thermal detection is reviewed (Chapter 9). Telemetry
compliments thermal census by facilitating detailed examination of
movements, behavior, and feeding preferences of instrumented individuals (Chapter 10).

Applications to aquatic environments are treated in Chapters 11-13.
Investigation of interactions of light in the surface layers of the sea
and its significance for delineating primary productivity (Chapter 11)
is followed by a study of shallow water benthic communities with color
photography (Chapter 12). The environment of Hudson Bay is partially
revealed by seasonal changes in the surface water temperatures (Chapter 13). Chapter 14 summarizes some of the advantages and techniques
of multispectral sensing for ecological applications. A selected bibliography of literature pertinent to remote sensing for ecological and natural
resource purposes is provided to guide additional reading.

Ecological Potentials in Spectral Signature Analysis[1]

R. M. HOFFER AND C. J. JOHANNSEN

INTRODUCTION

FOR many years, aerial photographs have been used by ecologists to map plant communities, identify species, and study changes that take place due to normal plant succession or such influences as fire or man. In the past few years increased use of color and color infrared photography has placed greater emphasis upon the spectral differences among plant species and the spectral changes that occur because of maturity or stress conditions. Color photography allows qualitative comparisons between the colors on the photos and the spectral characteristics of the material, since our eyes and the film are both sensitive to the same portion of the spectrum. The interpretation of photography in the infrared wavelengths, however, is more difficult. Knipling (1967) has pointed out that because color infrared film is sensitive to both the visible and infrared wavelengths it can be erroneously interpreted. In some cases, a change in the red tone of vegetation on the color infrared film has been interpreted as a change in plant reflectance in the infrared wavelengths, although in fact the difference was due to a change in plant reflectance in the visible wavelengths, and there was no change in the infrared reflectance of the plants. We have observed striking examples of this type of situation in which yellowing vegetation and healthy green vegetation reflected similarly on black and white infrared film (sensitive only to the $0.7 - 0.9\mu$ wavelengths), but marked differences were seen on the color infrared film due to spectral changes in the visible rather than the infrared wavelengths. This was also substantiated by spectrophotometric measurements of plant leaves.

Relatively recent developments in remote sensing systems allow us

1. Agricultural Experiment Station Journal No. 3479. This research was supported jointly by the U. S. Department of Agriculture and the National Aeronautics and Space Administration.

to record reflected or emitted energy in wavelengths far outside those
to which photographic emulsions are sensitive. As the potentials for these
systems become better known, we need to understand more fully the
factors affecting reflectance and emittance from various plant and soil
materials so that we may better predict the conditions under which we
would expect to observe changes in the spectral characteristics of these
materials. For example, assume that a certain disease condition causes a
stress condition in a plant that is known to affect the internal cell structure
of the leaves. We must be able to predict whether or not this condition
of stress is accompanied by a change in reflectance and emittance. We
must also become familiar with the manner and portion of the spectrum
in which the change will occur, so that we can then predict the potential
for using remote sensing systems to locate and map areas where this
disease condition exists. To do this research properly requires an integrated program of laboratory, field, and aircraft experiments. Without
such a program, however, we will find ourselves continually dealing
with new problems by empirical methods.

FUNDAMENTALS OF LEAF REFLECTANCE

A brief review of the optical portion of the spectrum, as shown in
Figure 1, indicates that up to approximately 3µ wavelengths, we are
concerned with reflective phenomena, due to the range of response of
solar energy. Suits (1960) points out that, "Solar reflectant power is
decreased with increasing wavelengths until the radiation emitted by
the object is dominant. The crossover point where the emitted radiation
becomes dominant over the reflected radiation is at approximately 3µ."

In the reflective wavelengths below 3µ, there are three major
areas:

1. ultraviolet wavelengths—approximately 0.3 to 0.38µ
2. visible wavelengths—approximately 0.38 to 0.72µ
3. reflective infrared—approximately 0.72 to 3µ

Energy recorded in the 3 to 15µ portion of the spectrum is mostly due
to emission from an object, which is a function of the true temperature and the emissivity of the object.

Figure 2 shows the characteristic spectral reflectance of a green
leaf and indicates that the 0.4 to 2.6µ portion of the spectrum can
be roughly divided into three areas. First is the visible wavelength
region in which plant pigments, especially the chlorophylls, dominate
the spectral response of plants. Second is the region from approximately 0.72 to 1.3µ where there is very little absorption by a leaf, and
therefore most of the energy impinging upon the leaf must be either
transmitted or reflected. The third region is that of water absorption,
extending from about 1.3 to 3µ. Figure 2 also indicates that a normal
green, healthy leaf will have four primary absorption bands. Two of
these are in the visible wavelengths and are caused by chlorophyll
absorption; one at approximately 0.45µ and one at approximately 0.65µ.

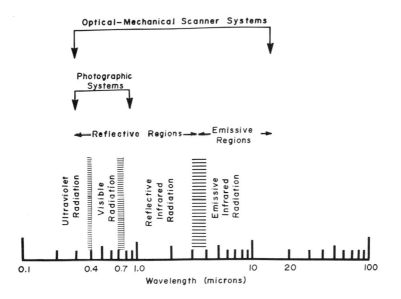

Figure 1. A Portion of the Electromagnetic Spectrum.

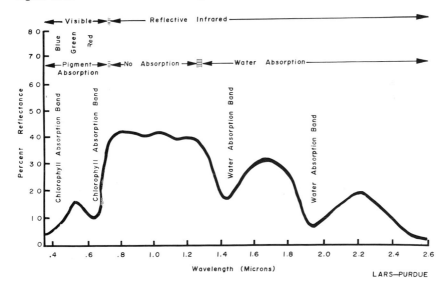

Figure 2. Three Primary Regions of Response in Relation to Leaf Reflectance.

Water absorption accounts for the strong decrease in reflectance in the two remaining wavelength bands, located at approximately 1.45 and 1.95μ.

LABORATORY STUDIES

Methods and Materials

In the summer of 1966, over 2300 spectra of various plant and soil samples were obtained at the Laboratory for Agricultural Remote Sensing. A Beckman DK-2 spectrophotometer was used, and data were obtained in the 0.5 - 2.6μ portion of the spectrum. A datex encoder and punch card unit enabled the reflectance to be automatically recorded on computer punch cards at 10mμ intervals, with an accuracy to the nearest 0.1% reflectance. Calibration curves at 0 and 100% reflectance were also obtained, and all data were then normalized to these calibration curves. This procedure corrected possible changes in the MgO standards, as well as any variations in instrument settings.

Most of the leaf data collected consisted of reflectance measurements on the ventral (top) side of the leaves. Some transmittance data were also obtained, which allowed determination of leaf absorptance using the relationship:

Absorptance = Incident Radiation — (Reflectance + Transmittance)

Cross sections of the leaves were obtained on about 10% of the samples, using the microtome techniques outlined by Jensen (1962). Moisture content of the leaves was also determined, using the formula:

$$\text{Moisture Content (in \%)} = 100 \left(\frac{\text{Fresh Leaf Weight} - \text{Dry Leaf Weight}}{\text{Fresh Leaf Weight}} \right)$$

Effect of Pigmentation on Leaf Reflectance

Leaf pigmentation can cause marked differences in spectral response in the visible wavelengths, as illustrated in Figure 3a. The white Coleus leaf without any apparent pigmentation has a very high level of reflectance throughout the 0.5 - 0.9μ region. The leaf dominated by chlorophyll pigmentation shows the characteristic curve for a green leaf, with relatively low reflectance at 0.5μ, a peak in the green at approximately 0.55μ, low reflectance again in the red at about 0.65μ, and then the usual sharp increase at about 0.7μ to the reflective infrared wavelengths. A red leaf had a low reflectance throughout the blue and green portions of the spectrum, then a marked increase and very high level of reflectance throughout the red and near infrared wavelengths. A deep reddish-purple leaf had a relatively low level of reflectance throughout the visible region, and then a sharp rise which coincides with that of the green leaf.

Figure 3b extends the wavelength band of interest for coleus leaves to 2.6μ. There is very little difference in reflectance among these four leaves throughout the reflective infrared wavelengths, in spite of the marked differences in the visible wavelengths caused by the pigmentation.

Figure 3c shows distinct differences in reflectance. Reflectance spectra were obtained of two silver maple (*Acer saccharinum* L) leaves

Figure 3. Pigmentation Effects on the Reflectance of Leaves of Different Plant Species.

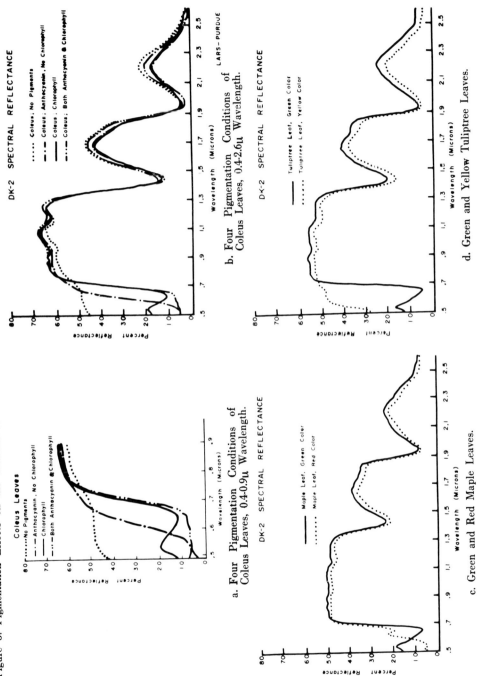

b. Four Pigmentation Conditions of Coleus Leaves, 0.4-2.6μ Wavelength.

d. Green and Yellow Tuliptree Leaves.

a. Four Pigmentation Conditions of Coleus Leaves, 0.4-0.9μ Wavelength.

c. Green and Red Maple Leaves.

in the fall, and are typical of several obtained for leaves at this stage of senescence. One of these leaves was a normal green, the other a brilliant red. The red coloration was caused by the presence of anthocyanins, which are often produced in maple trees in the fall after chlorophyll production has ceased. As was the situation with the red Coleus leaf, the red maple leaf has a relatively low reflectance at wavelengths below about 0.60μ. In this case, however, there was a decrease in reflectance at approximately 0.66μ in the red chlorophyll absorption band, thus indicating the presence of some minor amounts of chlorophyll. The generally high reflectance throughout the 0.62 - 0.72μ portion of the spectrum dominates, thus giving the leaf its brilliant red coloration, so familiar in the fall of the year.

Spectra for leaves of tuliptree (*Liriodendron tulipifera* L) are shown in Figure 3d. Both leaves were quite succulent and at normal high moisture contents. However, one leaf was a normal, deep green color while the other was a bright yellow. This yellow coloration was caused by the normal autumn breakdown of the chlorophylls, which were not reformed, revealing the presence of the carotenes and xanthophylls. These carotenoid pigments were present before the chlorophyll breakdown, but were masked by the chlorophylls. These spectra (Fig. 3d) show the usual curve for a green leaf, but the yellow leaf had a very sharp increase in reflectance starting at 0.50μ and continued high reflectance throughout the green and red portions of the visible spectrum. In the infrared wavelengths, the yellow leaf had 2-3% lower reflectance than the green leaf. This could cause a somewhat darker tone on black and white infrared film.

Effect of Moisture Content on Leaf Reflectance

Reflectance of leaves in the 1.3 to 2.6μ portion of the spectrum should be examined with particular attention to the water content of the leaves. Forsythe and Christison (1930) showed the absorption for one millimeter thickness of water in these wavelengths. Figure 4 indicates the close relationship between water absorption and reflectance for a healthy, turgid green leaf. In wavelengths where water absorption is high, leaf reflectance is low. This is most apparent in the primary water absorption bands centered at 1.45 and 1.95μ. There are also slight increases in water absorption at approximately 0.96 and 1.2 microns. These minor water absorption bands cause slight decreases in reflectance of the leaf. However, because water absorption is generally very low in the 0.7 and 1.3μ band, its influence upon leaf reflectance is minor.

The influence of the drying of leaf tissue and the marked spectral changes that take place in the water absorption region are shown in Figure 5a. These curves were averaged for groups of corn leaves at four different levels of moisture content. Marked increases in reflectance with decreasing moisture content were observed throughout the 0.5 - 2.6μ

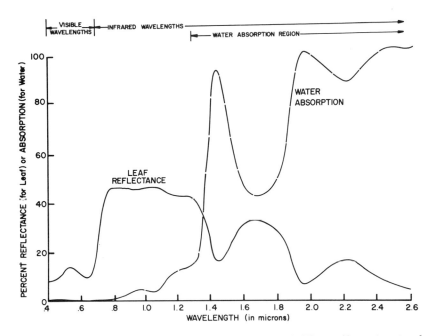

Figure 4. Relationship between Leaf Reflectance and Water Absorption in the 0.4-2.6μ Wavelength Region.

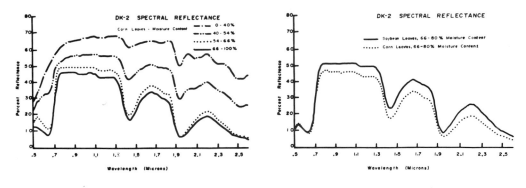

a. Corn Leaves at Four Moisture Levels.   b. Corn and Soybean Leaves at the Same Moisture Level.

Figure 5. Effects of Differences in Moisture Content on Leaf Reflectance Spectra.

region. The curve for corn leaf samples in the 4-40% moisture content range approached 70% reflectance in the unabsorbed wavelengths and decreased to about 42% at 2.50μ. Only slight decreases were seen in the primary water absorption bands. These leaves lacked chlorophyll and therefore had no absorption in the red chlorophyll absorption band in the visible wavelengths. The curve for the 40-54% moisture content range shows a 10-18% decrease in reflectance throughout the reflective infrared wavelengths. The primary water absorption bands are quite evident, and the general shape of the curve in the water absorption region from 1.3 - 2.6μ resembles that of normal succulent leaves more than does the curve for 0-040% moisture content. There was a slight decrease in reflectance at 0.64 - 0.66μ, indicating that there was a small amount of chlorophyll present.

The two remaining curves in Figure 5a (54-66% and 66-76% moisture content) are generally similar. Reflectance differences in the primary water absorption bands are from 0-3%, although in the regions between these water absorption bands there is a 3-4% difference in reflectance. In the unabsorbed region, there is a fairly consistent difference of 3-4%. Examination of leaf cross-sections indicated that this difference is not due to water absorption as much as to the structural changes taking place in the leaf that are accompanied by the loss in moisture content. The 54-66% moisture content curve has a higher response in the green-visible wavelengths due to a lighter color, as compared to the curve for leaves having 66-76% moisture content.

These curves indicate that reflectance measurements are strongly influenced by the moisture content of the leaves, particularly but not exclusively, in the 1.3 - 2.6μ region. Changes in the leaf structure and pigmentation accompanying the changes in moisture content also have strong effects on reflectance. The observed differences in reflectance as related to moisture content of leaves indicate a possible potential for remotely detecting moisture stress in plants using bands in the 1.3 - 2.6μ portion of the spectrum.

Reflectance of leaves from different species but in the same moisture content range may be significantly different. Averaged curves for leaf samples from 66-80% moisture content are shown in Figure 5b for 308 soybean leaves and 382 samples of corn leaves. The curves are very similar throughout the visible wavelengths. However, in the entire reflective infrared region from 0.72 to 2.6μ, the soybean leaves have a higher reflectance than the corn leaves. This is thought to be caused primarily by the structural differences between monocotyledonous and dicotyledonous leaves; the dorsi-ventral structure of the dicotyledonous leaves usually resulting in higher reflectance. This was found for the leaves of several plant species examined in detail (Sinclair 1968). Reflectance and transmission spectra for both monocotyledonous and dicotyledonous leaves indicate that very little energy is absorbed in these 0.72 - 1.3μ wavelengths by either type of leaf structure.

Spectra for the leaves of three species of trees are shown in Figure 6. Although all three spectra have approximately the same general shape, characteristic of green vegetation, there are significant differences in amplitude of reflectance in certain wavelength bands. These relative differences in reflectance in the various wavelength bands are significant in pattern recognition because they help identify unknown spectra through correlation among discrete wavelength bands. Most evident in these spectra are the marked differences in reflectances throughout the unabsorbed wavelengths, the tuliptree having about 5% higher reflectance than the silver maple, which in turn has about 5% higher reflectance than the American elm. In the visible wavelengths, the peak in the green shows highest reflectance from the maple leaf, and lowest reflectance from the elm. A crossover takes place in the red chlorophyll absorption band, however, and the tuliptree reflects the least in this wavelength band. The maple still has high reflectance. In the 1.9 - 2.6μ region, the elm has a somewhat higher reflectance than either the maple or tuliptree, which are quite similar in reflectance.

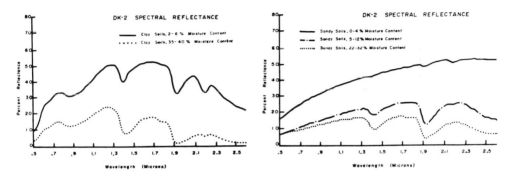

a. Clay Soils at Two Moisture Levels.

b. Sandy Soils at Three Moisture Levels.

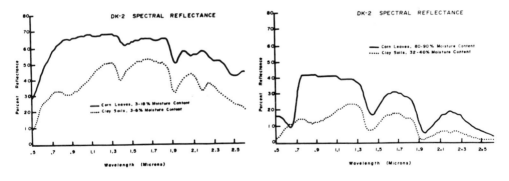

c. Corn Leaves and Clay Soils at Low Moisture Contents.

d. Corn Leaves and Clay Soils at High Moisture Contents.

Figure 6. Relationship between Reflectance and Moisture Content for Two Soil Types and Corn Leaves.

Effect of Moisture Content on Soil Reflectance

Spectral measurements of over 250 soil samples were obtained. Ten different soil textures, four drainage profiles and 3 major soil horizons were represented in these samples. Spectra of sandy and clay soils were selected to show the effect of moisture on soil reflectance. The mean spectral curves for the clay soils at two different moisture levels and sandy soils at three different moisture levels are shown in Figures 7a and 7b respectively. The curves for both soil categories show a very large decrease in reflectance with an increase in moisture. Curves for the clay soils maintain the same spectral shape, but the sandy soils exhibit a marked change in the shape of the spectral curve with a change in moisture. The water absorption bands at approximately 1.45 and 1.95μ become pronounced for sandy soils with a moisture content of over 4%. The authors noted that in pure samples of bentonite, muscovite, and kaolinate clays at only 0.1% moisture content still had strong water absorption bands still existed. This evidence, together with the lack of change in the characteristic spectral curves for soil at low moisture contents seems to indicate that bound water may be exerting an influence on the reflectance. The bound water in sandy soils is very low compared to clay soils. Thus the sandy soils at low moisture contents produce a rather flat, uniform reflectance curve throughout the reflective infrared wavelengths, without the decrease in reflectance in the strong water absorption bands seen for the clay soils at low moisture contents.

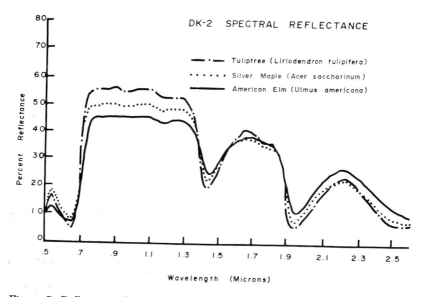

Figure 7. Reflectance Spectra for Leaves of Three Tree Species.

The ability to delineate soil types and to survey the soil moisture content would be of great interest to ecologists, soil scientists, and others. The data just presented represent reflectance from the soil surface. As noted, the reflectance is greatly reduced with an increase in moisture content of the soil surface. It is well known that soil tends to dry on the surface, forming a thin dry crust. This crust can develop within a few hours after a rain (Mannering 1967). Therefore, reflective measurements using remote sensing devices could show that the soil appears to be dry, whereas the soil profile might be very wet. This indicates a possible limitation in the utility of reflectance data alone. However, the use of the emissive wavelengths, perhaps in combination with the reflective wavelengths, may allow soil moisture information to be obtained by remote sensing techniques.

Comparisons of corn leaf reflectance at low moisture (0-20%) with reflectance of clay soils at low moisture (2-6%) and reflectance of corn leaves at high moisture (80-90%) with clay soils at high moisture (22-40%) are shown in Figure 7c and 7d respectively. The overall shapes of the low moisture curves are somewhat similar. The corn samples generally have a 10-15% higher reflectance in the water absorption regions. A characteristic decrease in reflectance for these clay soils at about $0.80 - 0.90\mu$ results in a difference in reflectance of over 30%. Also, the characteristic decrease in reflectance for these clay soils at 2.20μ is not observed in the curve for dry corn leaves.

In the comparison of high moisture clay and leaf samples at high moisture contents, it should be noted that the clay soils were completely saturated at 32-40% moisture, and the corn samples at 80-90% moisture content were also at a maximum moisture level. Even though the corn has a higher percentage moisture content than the clay, because of differences in these materials and in the total amount of water present, the clay has a lower reflectance in the water absorption region. Again, the similarity in shape of the curves in the water absorption region $(1.3 - 2.6\mu)$ is striking. Of course, the clay does not have the characteristic "green vegetation" reflectance in the visible wavelength bands, or the sharp rise at about 0.7μ. In the $0.72 - 1.0\mu$ region, the clay soils have about 30% lower reflectance than the corn leaves. This difference would explain the high response of green vegetation on black and white infrared film compared to the low response generally exhibited by soils.

Summary

With the use of these leaf and soil spectra, we have attempted to show the utility of laboratory spectral reflectance data for interpretation purposes. The portion of the spectrum where the greatest differences in reflectance among various plant and soil materials occur can be determined, and the effects of differences in moisture content, pigmentation, and internal leaf structure can be studied. We believe that laboratory studies are a very necessary step in developing an understanding of energy interactions with plant and soil materials. However, it must be stressed that a capability for differentiating plant or soil materials on the basis of laboratory spectra does not mean that the same results will be obtained in the field. Laboratory spectra were obtained on very small plant or soil areas normally oriented at an angle perpendicular or nearly perpendicular to the incident beam of energy. Such spectra are not necessarily comparable to the spectral response measured remotely as from an aircraft. For example, aircraft spectral data of a particular plant species may represent a mixture of target and background including leaves oriented at many different angles; some leaves are green and succulent, some brown and dry; shadow areas are present; some soil may be showing (with accompanying differences in response due to soil type, surface texture and moisture conditions etc.) (Hoffer 1967). In spite of such potential variations in the aircraft data, there exists a capability to use remote sensing spectral data to identify and map various plant and soil materials. The next section discusses one example of automatic classification and mapping of three basic cover types.

AIRCRAFT SPECTRAL STUDIES TO AUTOMATICALLY IDENTIFY COVER TYPE

In many ecological and agricultural situations there is a need to map various cover types over large geographic areas and at relatively frequent intervals. Schneider (1966) pointed out that ecological conditions in the Florida Everglades are changing very rapidly and that only a small portion of the entire area had been adequately mapped. In many agricultural situations there is also a need for mapping crop species and soil conditions over large areas. Such conditions may change rapidly, and are quite different from season to season and from year to year. Remote sensing offers an excellent method for covering large geographic areas in a relatively short period of time. However, to be able to reduce large quantities of remote sensing data to useful information in a timely manner, there is a necessity for automating the identification and mapping process. Methods have been developed whereby optical-mechanical scanners aboard aircraft can be flown over geographic areas of interest to record spectral data simultaneously in each of several discrete wavelength bands. In our research program at Purdue University, pattern recognition techniques are being applied to multispectral data obtained from aircraft altitudes, thereby permitting the automatic identification and mapping of various cover types and crop species. Although this research has been underway only since 1966, it appears that the application of pattern recognition techniques to crop species and cover type identification offers much promise (Baumgardner *et al.* 1967, Hoffer and Landgrebe 1965, LARS 1967 and LARS 1968).

Briefly, to apply pattern recognition techniques to such multispectral response data, the radiance (reflectance or emittance) of small areas on the ground is measured by the optical-mechanical scanner simultaneously in several wavelength bands. By ordering the measured response values for each of the wavelength bands used, one then obtains a measurement vector for each cover type, species, or other surficial material of interest. On the basis of these vectors, the pattern recognition categorizer is programmed to determine differences among the various materials. At this point, the categorizer is trained, using these known target materials, and then the measurement vectors of unknown target materials, or test samples, can be examined and classified.

In one recent experiment, multispectral scanner data were used in a classification of green vegetation, bare soil, and water for an area in central Indiana. The results of this classification are shown in Figure 8. On the left is an aerial photo mosaic of this area. In the center is a double-width computer printout of the same area. The computer printed a G for all areas identified automatically as green vegetation. These areas show up as a dark tone on the printout. Bare soil areas were printed out as a dash, and show up as light toned areas. The water was identified by a slash and is a somewhat intermediate tone on this computer printout. On the right is a computer printout of only the areas identified as water. Many of the points identified as water appear rather small and scattered, but are actually ponded areas and water in drainage ditches. It is of interest to note that there were several instances in which water was correctly identified automatically but which had previously been overlooked on the aerial photographs. Several of these areas of water were difficult to see on the aerial photos because of overhanging tree branches or a lack of distinctive tonal variations between the water and other materials in the vicinity.

The multispectral scanner data used were obtained at an altitude of 3,200 feet by the Institute of Science and Technology, University of Michigan. The automatic classification shown in Figure 8 was done at the Laboratory for Agricultural Remote Sensing, Purdue University, using the following four wavelength bands: $0.48 - 0.50\mu$, $0.58 - 0.62\mu$, $0.66 - 0.72\mu$, and $0.80 - 1.0\mu$. It should also be pointed out that these black and white aerial photos were obtained by the Indiana State Highway Commission ten days before the multi-spectral scanner data were collected, so that in several places ground conditions at the time of the scanner flight and at the time the photos were obtained were quite different.

To quantitatively check these classification results a total of 107 test sample areas were selected at random from the computer data. Based upon study of these photos and of ground truth data collected at the time of the scanner flight mission, 48 of these test areas were determined to belong to the green vegetation category, 42 as soil and 17 as water. Then the automatic classification of sample points within each test area was examined. In 96 of the test areas, a 97% or higher classification accuracy was obtained, and in all except one of the 107 test areas, a correct classification of over 83% of the sample points had been achieved. The average classification accuracy was 99.4% for green vegetation, 98.0% for soil and 96.7% for water. It is the inherent differences which exist in the spectral signatures of these three materials that explain this capability for automatic identification.

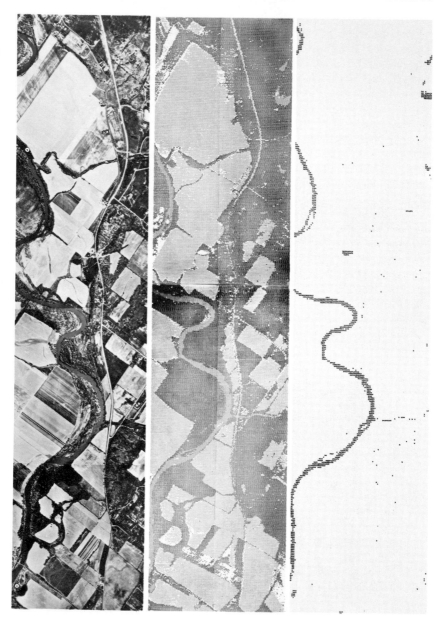

| Aerial Photograph | Computer Printout of Green Vegetation, Soil, and Water | Computer Printout of Water Only |

Figure 8. Aerial photograph and computer printouts showing automatic identification of green vegetation, soil, and water.

Summary

The significance of this procedure for applying pattern recognition techniques to multispectral scanner data is that it allows data to be obtained rapidly over very large geographic areas, followed by automatic processing of the data to a reasonable degree of accuracy for identification and mapping of various cover types. Such procedures would have great promise in situations such as exist in the Florida Everglades, where the data could be obtained and processed at rather frequent intervals. Optical-mechanical scanners do not have the all-weather capability of radar, but do have some advantages in terms of resolution, power requirements, and in data handling. Photography offers much better resolution than scanner data but covers only a narrow portion of the spectrum (0.4 to 0.9 μ) whereas the scanner capability extends from 0.3 to 15 μ. Optical-mechanical scanner data has another advantage in the ease with which the data can be quantified and processed by computer. Therefore, it would appear that an operational remote sensing system of the future should incorporate the unique advantages of several instrument systems including photographic sensors, optical-mechanical scanner, radar, and perhaps laser altimeters and other instruments. We must not limit ourselves to the advantages of one instrument system and overlook the obvious advantages of others which could and should be used together in remote sensing programs.

Acknowledgements

The Mobile Spectrophotometry Laboratory used in this study is owned by the Office of Naval Research and was loaned to Purdue University by Mr. Charles Olson, School of Natural Resources, University of Michigan, to whom appreciation is expressed.

The aircraft scanner and photographic data shown in Figure 8 were made available to LARS by the Airphoto Interpretation and Photogrammetry Laboratory, School of Civil Engineering, Purdue, University. These data were obtained for research sponsored by the Indiana State Highway Commission and the Bureau of Public Roads, U. S. Department of Transportation.

Mr. Roger Wyse and Mr. Thomas Sinclair assisted in spectral data collection. R. B. MacDonald, D. A. Landgrebe, T. L. Phillips. P. Swain, S. Kristof and others of the LARS staff are acknowledged for their parts in the handling and analysis of the multispectral scanner data. We also acknowledge J. Halsema for the photography.

REFERENCES

Baumgardner, M. F., R. M. Hoffer, C. J. Johannsen and C. H. Kozin. 1967. Contributions of automatic crop surveys to agricultural development. Paper presented at AIAA 4th Annual Meeting and Technical Display, Anaheim, California. October 23-27.

Forsythe, W. E. and F. L. Christison. 1930. The absorption of radiation from different sources by water and by body tissue. Jour. Optical Soc. Amer. 20:693-700.

Hoffer, R. M. 1967. Interpretation of remote multispectral imagery of agricultural crops. Laboratory for Agricultural Remote Sensing. Vol. No. 1. Purdue University. 36 pp.

Hoffer, R. M. and D. A. Landgrebe. 1965. Automatic processing of multispectral scanner data. Paper presented at 34th Annual Meeting of American Society of Photogrammetry, Washington, D. C.

Jensen, W. A. 1962. *Botanical Histochemistry - Principals and Properties.* W. H. Freeman and Co., San Francisco, California.

Knipling, E. B. 1967. Physical and physiological basis for differences in reflectance of healthy and diseased plants. Workshop on Infrared Color Photography in the Plant Sciences. March 2, 3. Florida Dept. Agriculture, Winter Haven, Florida.

Mannering, J. V. 1967. The relationship of some physical and chemical properties of soils to surface sealing. Unpublished Ph.D. thesis. Purdue University, Lafayette, Indiana.

Miller, E. V. 1953. *Within the Living Plant.* The Blakiston Company, Inc. New York.

Moss, R. A. and W. E. Loomis, 1952. Absorption spectra of leaves. I. The visible spectrum. Plant Physiology 37:370-377.

Myers, V. I., C. L. Wiegand, M. D. Heilman, and J. R. Thomas. 1966. Remote sensing in soil and water conservation research. Proc. Fourth Symp. Remote Sensing of Environment, Univ. Michigan, Ann Arbor. Pp.801-813.

Schneider, W. J. 1966. Water resources in the everglades. Photogr. Engr. 32:958-965.

Sinclair, T. R. 1968. Pathway of solar radiation through leaves. Unpublished M.S. thesis. Purdue University, Lafayette, Indiana.

Suits, G. H. 1960. The nature of infrared radiation and ways to photograph it. Photogr. Engr. 26:763-772.

Laboratory for Agricultural Remote Sensing 1967. Remote multispectral sensing in agriculture. Volume No. 2, (Annual Report) Research Bulletin No. 832, Agricultural Experiment Station, Purdue University. 75 pp.

Laboratory for Agricultural Remote Sensing. 1967. Remote multispectral sensing in agriculture. Volume No. 2, (Annual Report). Research Bulletin No. 844, Agricultural Experiment Station, Purdue University, 176 pp.

2

Leaf Reflectance and Image Formation on Color Infrared Film

Edward B. Knipling

INTRODUCTION

In recent years applied ecologists have found considerable use for false color aerial photography with Ektachrome Infrared Aero film (EKIR) for crop and vegetation studies and surveys (Carneggie 1968, Ciesla *et al.* 1967, Heller 1968, Johnson 1965, Meyer and French 1967, Norman 1967, Norman and Fritz 1965). Healthy plant foliage characteristically appears a bright red or magenta color with different species often distinguishable by varying shades, whereas unhealthy, damaged, or dying vegetation tends to deviate from the red color. The image color formation is dependent on reflected energy of the green and red portions of the visible spectrum as well as the near-infrared. The features of plant leaves basically responsible for their unique red color on the photographs are their relatively low level of visible reflectance and high level of infrared reflectance.

Because the film is commonly called an "infrared film," there has been a natural tendency by many workers to attribute all deviations from the red color of plants on EKIR photographs to a lack of or decline in infrared reflectance. This concept has been sustained by claims that the level of infrared reflectance of plant leaves is a sensitive indicator of the healthiness and physiological state of plants and that during initial stages of stress it decreases before visual changes or symptoms appear (Colwell 1967). However, more and more workers are reporting that most of the color differences of plants observed on the false color photography also can be found visually and with conventional photography (Benson and Sims 1967, Ciesla *et al.* 1967, Heller 1968, Knipling 1967). These findings suggest that changes or differences in the visible reflectance of leaves are perhaps as sensitive indicators of physiological stress in plants as are changes in infrared reflectance and also that they contribute significantly to the color differences of EKIR photographs.

17

Thus, in spite of the fact that EKIR film has found widepsread and successful use, there exist some misunderstandings and misconceptions about the process of color formation and about the way the reflectance properties of leaves change in the spectral sensitivity range of the film. The purpose of this discussion is to relate the characteristics of leaf reflectance with the properties of EKIR film, and the interaction between the two. By focusing attention on the way the colored imagery of plant foliage is produced on this film, a basis for the successful use of the film will result.

PROPERTIES AND MECHANISM OF LEAF REFLECTANCE

A typical reflectance spectrum of a plant leaf is shown in Figure 1. The percentage reflection is low in the visible region with a slight peak in the green portion at about 530 mμ. In contrast, the reflectance in the near-infrared is at a quite high level, about 60%, but it gradually decreases beyond 1200 mμ to a low value of at 2700 mμ with two intervening bands of low reflection at about 1450 and 1950 mμ. Beyond the ends of the wavelength range covered in Figure 1, (i.e. in the ultraviolet and the far-infrared), the leaf reflectance is low, generally 0 to 10% (Gates and Tantraporn 1952, Wong and Blevin 1967).

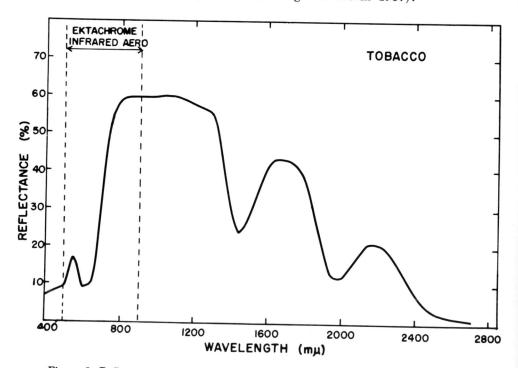

Figure 1. Reflectance spectrum of a tobacco leaf and the spectral sensitivity range of Ektachrome Infrared Aero film.

Since reflectance in Figure 1 is expressed in percent, the exact shape of the curve is not truly representative of the amount of energy reflected from the leaf because the intensity of incident solar energy is not uniform across the spectrum. In fact, when energy is expressed on the basis of the amount per wavelength interval, the incident solar energy peaks in the visible at about 475 to 500 mμ and decreases to a relatively low level in the near infrared. Thus, the shape of the incident energy spectrum is nearly the opposite of the leaf reflectance spectrum and therefore tends to compensate somewhat for the large difference in reflectance between the two regions. However, the infrared reflectance expressed on an energy basis is still higher than the visible reflectance.

The transmittance spectrum of a leaf is of the same general shape as the reflectance spectrum, that is low in the visible and high in the infrared (Gates et al. 1967, Sinclair et al. 1968). However, the absorption spectrum is the opposite or the inverse of the other two. All three spectra are closely interrelated with the strong absorption by compounds within a leaf indirectly accounting for regions of low reflectance and transmittance. In the infrared beyond 1200 mμ, especially at 1450 and 1950 mμ, the high absorption is by water (Allen and Richardson 1968, Gates et al. 1965). In the visible the high absorption is by chlorophyll (Mestre 1935, Rabideau et al. 1946, Shull 1929, Turrell et al. 1961, Willstatter and Stoll 1918). This is well illustrated by the low absorption of and the high visible reflectance from a leaf which lacks chlorophyll such as the white or albino portion of a variegated leaf (Billings and Morris 1951, Gates et al. 1965). In the near-infrared there is no absorption by chlorophyll, and since the level of reflectance here is about the same for both the white and green parts of the leaf, it can be concluded that neither the absence nor the presence of chlorophyll is responsible for the infrared reflectance. This is an important point, for the high level of infrared reflectance of leaves has sometimes erroneously been attributed to chlorophyll. Instead, the infrared reflectance is caused by the internal leaf structure.

Nearly all of the incident infrared solar energy penetrates the leaf cuticle and epidermis and is completely diffused and scattered inside the leaf, primarily at the cell walls where refractive index differences between air and hydrated cellulose occur (Gates et al. 1965, Gates 1967, Knipling 1967, Mestre 1935, Sinclair et al. 1968, Willstatter and Stoll 1918). The cellulose microfibrils, both within the cell wall and at the surface, probably account for the diffusing nature of the walls (Sinclair et al. 1968). Generally about 40 to 60% of the near-infrared radiation is scattered upward through the surface of incidence and constitutes what is called reflected radiation, whereas the remaining is scattered downward and is transmitted. The high level of visible reflectance from a white leaf indicates that the interaction of energy in this wavelength region with the leaf structure is not really different from the interaction of infrared energy. However, when chlorophyll is present, about 80 to 90% of the scattered visible energy is absorbed before it escapes the leaf.

The fact that chlorophyll absorbs most of the visible radiation energy and that water absorbs some of the infrared radiation is actually strong evidence for the internal reflectance mechanism, for the radiation must enter the leaf before it can be absorbed. Further evidence is given by the drastic reduction in the infrared reflectance as a leaf is infiltrated with water under vacuum (Gates *et al.* 1965, Mestre 1935, Willstatter and Stoll 1918). The water fills the air cavities and forms a continuous liquid phase medium throughout the leaf. The elimination of refractive index differences within the leaf increases direct transmittance at the expense of multiple scattering.

Much has been said and written about the importance of the air cavity-cell wall interfaces of the spongy mesophyll as the primary reflecting surfaces within the leaf. However, the necessity for large air cavities has probably been over-emphasized and it is doubtful if the spongy mesophyll region is really more important than the palisade mesophyll. Many small air cavities and fissures exist between adjacent palisade cells and the area of exposed cell walls probably is as large, and perhaps even larger in some cases, as in the spongy mesophyll which has larger air cavities and fewer cells.

Along with claims citing the importance of the spongy mesophyll in reflectance, the expression "collapse of the mesophyll" is often loosely used to explain and predict decreases in infrared reflectance. It is presumed that when leaves wilt and shrivel up during senescence and dehydration, some of the reflective interfaces are eliminated as the volume of internal air cavities is reduced. However, it is now well documented that the infrared reflectance of dehydrating leaves changes very little in the wilting range but increases with severe dehydration, especially in the water absorption bands (Allen and Richardson 1968, Sinclair *et al.* 1968, Thomas *et al.* 1966, Weber and Olson 1967). An example is shown in Figure 2. It also appears that in some cases the initial stages of disease and leaf senescence are accompanied by an *increase* in infrared reflectance (Keegan *et al.* 1956, Knipling 1967). Even though the volume of internal air space decreases as a leaf wilts, it is likely that the area of reflective interfaces actually increases as adjacent cells split apart and living cell contents shrink away from the interior cell walls. Also, there is evidence that the increased reflectance might be caused by the reorientation of the cell walls into the same plane as the leaf surfaces (Sinclair *et al.* 1968). The infrared reflectance eventually decreases in advanced stages of senescence, but at this stage it is more likely caused by an actual breakdown or deterioration of cell walls rather than by a reduction in the volume of internal air space.

COLOR FORMATION BY THE FILM

The spectral sensitivity range of Ektachrome Infrared Aero Film is 500 to 900 mμ, and it is the leaf reflectance properties in this region

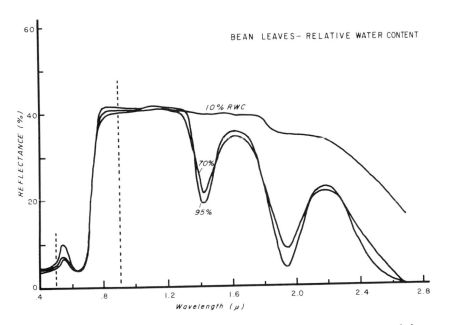

Figure 2. The effect of leaf dehydration on the spectral reflectance of bean leaves. The dashed lines delineate the spectral sensitivity range of Ektachrome Infrared Aero film.

(Figure 1) that need to be considered to relate leaf reflectance to color formation on the film. It should be emphasized that image formation on EKIR film is strictly a photographic process and is not dependent on emitted thermal energy. Thermal infrared sensing is a completely different process which operates at much longer wavelengths in the infrared, about 3 to 5μ and 8 to 14μ, where objects and surfaces at ambient environmental temperatures emit radiation energy with a peak intensity.

Figure 3 shows the sensitivity curves of the EKIR film emulsion (Fritz 1967, Tarkington and Sorem 1963) which consists of three layers. The top layer is the cyan-forming layer and is sensitive primarily to the near-infrared radiation, the middle layer is the yellow-forming layer and is sensitive to the green visible radiation and the bottom layer is the magenta-forming layer and is sensitive to the red visible radiation. Thus, two of the three layers are sensitive to visible radiation. In this sense the film is not strictly a true infrared film, but is more appropriately termed a visual-infrared film. The failure of some workers to recognize that changes in reflectance in the visible part of the spectrum are as important in determining film colors as are changes in the infrared, is the source of some of the misunderstanding about this film.

Because of reversal processing during the film development procedure, the dye of an emulsion layer does not form if the layer was exposed

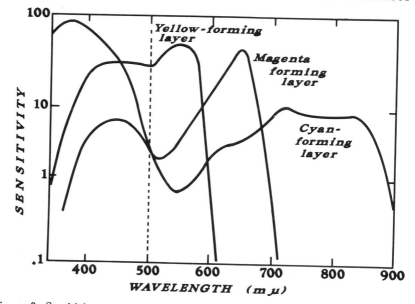

Figure 3. Sensitivity curves of the three emulsion layers of Ektachrome Infrared
Aero film. A yellow (minus blue) filter over the camera lens prevents exposure
by wavelengths below 500 mμ.

to the radiation to which it is sensitive. The way this process leads to
the photographic image color is illustrated in Figure 4. The arrows
at the top of the diagram represent reflected energy from the specific
regions of the visible and near-infrared spectrum. All three emulsions
of the film are sensitive to visible radiation below 500 mμ (Figure 3)
but these wavelengths are eliminated by using a yellow or minus-blue
filter over the camera lens. Thus objects reflecting only blue radiation
do not sensitize the emulsion and the dyes in all three layers form.
Each layer acts as a discrete color transparency or subtractive filter and
when all three are superimposed, no light can be transmitted. Thus,
the image of blue-reflecting objects appears black.

An object that reflects only infrared energy will expose the cyan
layer, leaving the yellow and magenta layer which combine in a subtractive mixture to form a red image when viewed by transmitted light.
Plant leaves reflect a significant amount of green energy and partially
expose the yellow layer in addition to nearly complete exposure of the
cyan layer by infrared. Therefore the magenta layer and part of the
yellow layer are left intact after the film development. The color of
vegetation thus varies from magenta to red.

An early version of Ektachrome Infrared Aero film originally was
developed by Eastman Kodak during World War II for use by the
military to distinguish green vegetation from green-painted objects
and dyed camouflage netting that did not have high levels of infrared

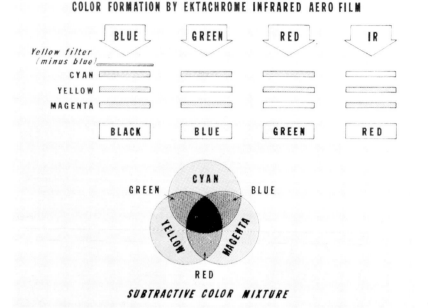

COLOR FORMATION BY EKTACHROME INFRARED AERO FILM

SUBTRACTIVE COLOR MIXTURE

Figure 4. Color formation by Ektachrome Infrared Aero film. Blue radiation does not expose the emulsion layers because of the yellow filter. All three dyes develop and combine in a subtractive mixture to form a black image. Green, red, and infrared radiation expose the yellow, magenta, and cyan layers respectively. When any one layer is exposed, the dye in that layer does not develop and the two remaining layers combine in a subtractive mixture to form the image color. When reflected energy from a surface exposes two layers, the image color becomes that of the single layer not exposed.

reflectance (Clark 1967). From the color formation chart it can be seen that such painted and dyed objects would appear blue on the photographs in contrast to the red of vegetation. Because of this use, the film became known as camouflage detection or C D film. Actually, it appears that this name is partially responsible for some of the present-day misconceptions about the film's capabilities and about the reflectance properties of leaves. The word camouflage has tended to imply the use of cut vegetation for concealing objects or positions, and the claim naturally evolved that the basis for the successful use of the film was that freshly-cut vegetation loses its high level of infrared reflectance. However, as shown in Figure 2, the infrared reflectance of dehydrating, cut leaves changes very little in the sensitivity range of the film.

Experimental Results

Some specific examples of reflectance curves of healthy and senescing leaves and the descriptions of resulting image colors on EKIR photographs will now be discussed. Figure 5 shows the reflectance of healthy (green and mature) and senescing (yellowing) African violet leaves.

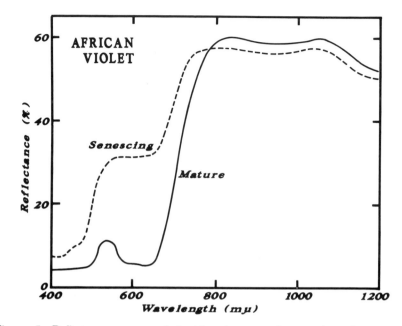

Figure 5. Reflectance spectra of healthy (green and mature) and senescing (yellowing) African Violet leaves.

The infrared reflectance of the senescing leaves decreased slightly but the most significant change was the increase in the visible reflectance. In an EKIR photograph the senescing leaves appear a light pink color, and this deviation from the characteristic red color of healthy green leaves is caused not by the change in infrared reflectance but by the increase in visible reflectance. In fact, by referring to the color formation chart (Figure 4) it can be seen that a high level of reflectance in the visible as well as the infrared is required to produce the color of the senescing leaves.

Figure 6 shows the reflectance curves of the red and green portions of a maple leaf during the fall coloration. The infrared reflectance of the red portion of the leaf increased slightly but the major change was the loss of the green peak and the increase of the red reflectance within the visible spectrum. When the reflected energy from the red parts of the leaf is exposed to EKIR film, the magenta and cyan layers are sensitized by the red and infrared reflectance respectively, leaving only the yellow dye intact. Thus, the image color is yellow.

Figure 7 shows the reflectance curves of four beech leaves (designated by numbers 1, 3, 5, and 7) in various stages of fall coloration and senescence. Leaf 1 is a healthy green leaf, whereas leaf 7 is quite dead and dry and has turned dark brown. The latter leaf has lost the high shoulder of infrared reflectance at 700 mμ, but the level remains high

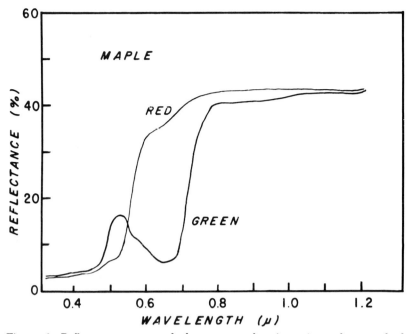

Figure 6. Reflectance spectra of the green and red portions of a maple leaf during fall coloration.

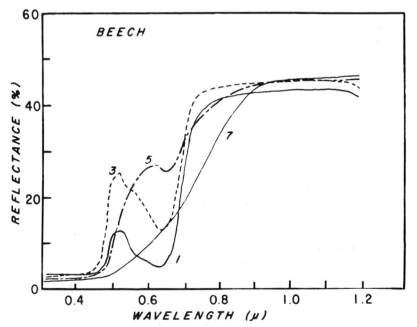

Figure 7. Reflectance spectra of a healthy beech leaf (No. 1) and senescing beech leaves (Nos. 3, 5, and 7) during fall coloration.

beyond 800 mμ. However, since the green reflectance has also decreased to a low level and the red reflectance has increased, the color infrared photograph of this leaf is dark orange. In leaves *3* and *5*, the early and intermediate stages of senescence, both the infrared and visible reflectance increased. The color infrared photographs of these leaves are pink and yellow respectively.

These examples support the argument that changes in the infrared reflectance of leaves, especially declines in reflectance, are not always sensitive indicators of anomalies or stress conditions in plants. Changes in the visible reflectance are in many cases more responsive indicators because of the sensitivity of chlorophyll to metabolic disturbances. As the chlorophyll deteriorates and absorbs less efficiently, the visible reflectance increases. Furthermore, the examples indicate that deviations from the red color of plant foliage on the false color infrared photography are not always caused primarily by a decline in infrared reflectance, but in many common cases by an increase in the visible reflectance. However, since the film integrates and records the reflectance differences in both regions of the spectrum, it is probably a better indicator of stress conditions than are the reflectance characteristics in either region alone. This is an important feature of the film and undoubtedly is partially responsible for its successful use.

DISCUSSION

One can generalize that foliage colors of red, pink, white, yellow, orange and brown in EKIR photographs are all caused by relatively high levels of infrared reflectance and by variable levels of visible reflectance. Vegetation that appears blue, green, or black indicates a low infrared reflectance. However, in these cases, the foliage is generally sparse, and the blue or green color results largely from the reflectance of limbs, branches, or soil and non-living ground cover exposed through openings in the vegetation cover or through the nearly bare crowns. Thus many of the very striking color differences on EKIR aerial photographs, i.e., red versus blue or green, that at first appear to be associated with differences in foliage vigor are actually caused by foliage and non-foliage surfaces respectively.

In addition to non-foliage reflective surfaces, shadows form pockets of low reflectance and generally cause a dark mottled appearance on the photographs. This effect is quite variable for it is a function of the illumination angle and the density and orientation of the leaves as well as the overall geometry and structure of the crop or plant community.

From the foregoing discussion, it can be seen that the reflectance characteristics of an entire plant or community are not always represented by single leaf reflection data. They are qualitatively the same, but differ quantitatively, with the percentage reflectance from a vegetation cover being generally lower than from a single leaf. For example, the level of visible reflectance from a broad-leaved forest canopy is about

40% of that from a single leaf and the infrared reflectance is about 70% of that from a single leaf (Steiner and Gutermann 1966). The smaller reduction in the infrared reflectance is because of a compensating factor. Much of the incident infrared energy which is transmitted through the uppermost leaves of a canopy is reflected from lower leaves and retransmitted up through the upper leaves to enhance their reflectance (Allen and Richardson 1968, Myers *et al.* 1966). This enhancement effect is nicely illustrated on close-up, oblique EKIR photographs of plants with overlapping leaves, for example a clump of grass. Where green leaves overlap, the total infrared reflectance is higher and the resulting image color is a brighter red than for only a single leaf.

Many of the color differences of vegetation on EKIR photographs, particularly the subtle shades of red, can be traced to variations in foliage area, density, and orientation rather than to differences or changes in the reflectance properties of individual leaves. Thus, it seems likely that herein lies part of the key to the basis for the successful use of the film in assessing plant vigor. A primary consequence of limiting environmental conditions and of disease, damage, and physiological stress in plants is a reduction in growth and thus in leaf area and foliage density. Concurrent with these changes are an increase in non-foliage reflective surfaces and shadows. The total amount of energy reflected from such areas would be much less than from healthy and vigorously growing vegetation even though the reflectance characteristics of individual leaves may not differ greatly.

In addition to differences in leaf reflectance, there are intrinsic emulsion and photograph characteristics that contribute greatly to the successful features of Ektachrome Infrared Aero film. The sensitivities of the emulsion layers and the color balance of the dyes enhance and amplify color differences that on conventional color photography may be questionable or overlooked. Also, the resulting red colors associated with plant foliage are brighter and lighter in tone than the normally dark shades of green on conventional color photography, and they generally are more pleasing to the human eye. Even though some experienced photo-interpreters have found that EKIR photographs do not reveal differences that are completely invisible to the human eye or that are not found on conventional color photography, the differences generally can be more readily spotted on the color infrared photographs and they lend reliability to the interpretation.

SUMMARY

False color aerial photography with Ektachrome Infarared Aero film has been found useful for distinguishing vegetation types and assessing plant vigor. However, considerable misunderstanding exists about the process of color formation on this visible—and infrared—sensitive film and about the way changes in leaf reflectance in these spectral regions account for differences in the color imagery of plant foliage.

Healthy green leaves have a low visible and high infrared reflectance and characteristically appear bright red on color infrared photographs. Physiological disturbances to plants generally are accompanied by increases in the visible reflectance but the direction of change in infrared reflectance is quite variable. Thus, deviations from the red color on photographs are not always explained by a decline in infrared reflectance. Disease, damage, and physiological stresses in plants influence the extent of red coloration by changing the geometry and density of foliage as much as by changing the reflectance characteristics of individual leaves.

REFERENCES

Allen, W. A. and A. J. Richardson. 1968. Interaction of light with a plant canopy. Proc. Fifth Symp. Remote Sensing Environment. Univ. Michigan, Ann Arbor, pp. 219-232.

Benson, M. L and W. G. Sims. 1967. False-color film fails in practice. J. Forestry 65:904.

Billings, W. D. and R. J. Morris. 1951. Reflection of visible and infrared radiation from leaves of different ecological groups. Amer. J. Botany 38:327-331.

Carneggie, D. M. 1968. Applying remote sensing techniques for improving range resource inventories. Proc. Fifth Symp. on Remote Sensing of Environment. Univ. Michigan, Ann Arbor, pp. 373-385.

Ciesla, W. M., J. C. Bell, and J. W. Curlin. 1967. Color photos and the southern pine beetle. Photogr. Engr. 33:883-888.

Clark, W. 1967. Aerial photography by infrared: a few historical notes. Proc. Workshop on Infrared Color Photography in the Plant Sciences, Florida Dept. Agriculture, Winter Haven, Florida.

Colwell, R. N. 1967. Remote sensing as a means of determining ecological conditions. BioScience 17:444-449.

Fritz, N. L. 1967. Optimum methods for using infrared-sensitive color films. Photogr. Engr. 33:1128-1138.

Gates, D. M. and W. Tantraporn. 1952. The reflectivity of deciduous trees and herbaceous plants in the infrared to 25 microns. Science 115:613-616.

Gates, D. M., H. F. Keegan, J. C. Schleter, and V. R. Weidner. 1965. Spectral properties of plants. Applied Optics 4:11-20.

Gates, D M. 1967. Remote sensing for the biologist. BioScience 17:303-307.

Heller, R. C. 1968. Previsual detection of ponderosa pine trees dying from bark beetle attack, Proc. Fifth Symp. Remote Sensing Environment. Univ. Michigan, Ann Arbor, pp. 387-434.

Johnson, P. L. 1965. Monitoring radioactive contamination to vegetation. Photogr. Engr. 31:984-990.

Keegan, H. J., J. C. Schleter, W. A. Hall, and G. M. Haas. 1956. Spectrophotometric and colorimetric study of diseased and rust resisting cereal crops. National Bureau of Standards, Report 4591.

Knipling, E. B. 1967. Physical and physiological basis for differences in reflectance of healthy and diseased plants. Proc. workshop on infrared color photography in the plant sciences, Florida Dept. Agriculture. Winter Haven, Florida.

Mestre, H. 1935. The absorption of radiation by leaves and algae. Cold Spring Harbor Symposia, Quant. Biology 3:191-209.

Meyer, M. P. and D. W. French. 1967. Detection of diseased trees. Photogr. Engr. 33:1035-1040.

Myers, V. I., C. L. Wiegand, M. D. Heilman, and J. R. Thomas. 1966. Remote sensing in soil and water conservation research. Proc. Fourth Symposium Remote Sensing Environment, Univ. Michigan, Ann Arbor, pp. 801-813.

Norman, G. G., ed. 1967. Proc. workshop on infrared color photography in the plant sciences. Florida Dept. Agriculture. Winter Haven, Florida.

Norman, G. G. and N. L. Fritz. 1965. Infrared photography as an indicator disease and decline in citrus trees. Florida State Horitcultural Society 78:59-63.

Rabideau, G. S., C. S. French, and A. S. Holt. 1946. The absorption and reflection spectrum of leaves, chloroplast suspensions, and chloroplast fragments as measured in an Ulbricht sphere. Amer. J. Botany 33:769-777.

Shull, C. A. 1929. Spectropotometric study of reflection of light from leaf surfaces. Botan. Gaz. 87:583-607.

Sinclair, T. R., M. M. Schreiber and R. M. Hoffer. 1969. Pathway of solar radiation through leaves (In Press).

Steiner, D. and T. Gutermann. 1966. Russian data on spectral reflectance of vegetation, soil, and rock types. Final Technical Report, U. S. Army European Research Office, 232 pp.

Tarkington, R. G. and A. L. Sorem. 1963. Color and false-color films for aerial photography. Photogr. Engr. 29:88-95.

Thomas, J. R., V. I. Myers, M. D. Heilman, and C. L. Weigand. 1966. Factors affecting light reflectance of cotton, Proc. Fourth Symposium on Remote Sensing of Environment. Institute of Science and Technology, Univ. Michigan, Ann Arbor, pp. 305-312.

Turrell, F. M., J. R. Weber, and S. W. Austin. 1961. Chlorophyll content and reflection spectra of citrus leaves. Botan. Gaz. 123:10-15.

Weber, F. P. and C. E. Olson. 1967. Remote sensing implications of changes in physiologic structure and function of the tree seedlings under moisture stress. Annual Progress Report, Remote Sensing Applications in Forestry, Natural Resources Program, National Aeronautics and Space Administration.

Willstatter, R. and A. Stoll. 1918. Untersuchungr uber die Assimilation der Kohlensaure. Springer, Berlin.

Wong, C. L. and W. R. Blevin. 1967. Infrared reflectances of plant leaves. Aust. J. Biological Science 20:501-508.

3

Large-Scale Color Photography Reflects Changes in a Forest Community During a Spruce Budworm Epidemic*

R. C. ALDRICH AND R. C. HELLER

INTRODUCTION

WHEN spruce budworm (*Choristoneura fumiferana* Clem.) attacks a spruce-fir forest and populations rise to epidemic levels, the forest community may undergo changes. The best evidence of this change is the increasing number of dead and dying balsam fir (*Abies balsamea* [L.] Mill), favorite host of the budworm. As the stands gradually decline and growing space increases, there may be greater competition for newly available air space, soil, and water. New openings created in the forest bring solar radiation to understory vegetation and soil that have received little light for many years. The questions are, *What changes occurred in the forest community as a result of these factors?* and *Will large-scale color photography provide a useful measure of the changes?* This chapter reports some preliminary findings of changes in a forest community as interpreted from large-scale aerial color photographs.

The most recent outbreak of spruce budworm in northeastern Minnesota was detected on the Superior National Forest in 1954 (Bean and Waters 1961). The epidemic reached a peak in 1961 when more than one million acres were defoliated to some degree. By 1963 noticeable current defoliation had dwindled considerably, indicating a slowdown in the epidemic (Ryan and Batzer 1964). This change was brought about by the partial reduction of the susceptible spruce-fir type through mortalty.

The outbreak in Minnesota provided an opportunity in 1957 to make a comprehensive study of aerial survey techniques with four major

*Remote Sensing Project, Pacific Southwest Forest and Range Experiment Station, Forest Service, U. S. Department of Agriculture, Berkeley, California.

objectives: (1) to develop an operational survey technique to detect and map the extent of budworm defoliation, (2) to test large-scale color photography to assess damage, (3) to test the feasibility of identifying boreal tree species, and (4) to test the feasibility of color photographs for measuring changes in the forest community.

THE STUDY AREAS

Nine 2 x 5 chain[1] each one acre plots, were selected from a larger number originally used in a budworm-damage classification study. The nine plots had to have a fairly dense crown cover; at least 30 – 50% of the crown closure was occupied by balsam fir. They were selected from three defoliation categories (Heller and Schmiege 1962): (1) light defoliation with 0-4 larvae, (2) moderate to heavy defoliation with 10-25 larvae, and (3) severe defoliation with 5-15 larvae per 15-inch branch sample. Each category was replicated three times. Soil, site, and topography were not considered in selecting plots.

METHODS

Ground Investigations

Each plot was marked for aerial photography with *L* shaped white panels at each corner. Markers were placed in openings near the corner and referenced to the corner by distance and azimuth. Plot diagrams recorded the corner locations. Precise measurements between panels and to plot corners were helpful later in determining photographic scale and in locating the one-acre plot boundaries.

In 1958 and each year through 1966, counts of budworm larvae were made on each plot. The counts were taken in late May or early June when larvae were near maturity. Five dominant or codominant balsam firs were sampled on each plot. On each tree, two 15-inch branches were clipped from opposite sides of the crown at midcrown level. The 15-inch branch sampling technique was developed by L. C. Beckwith at the North Central Forest Experiment Station, St. Paul, Minnesota. The ten samples per plot were bagged and taken to a laboratory for counting. Larvae were counted for each sample tree and an average calculated for each plot.

Defoliation estimates were made in July near the end of the larval feeding period. They were based on the number of needles removed from current growth on one 15-inch-long branch sample taken from the midcrown of ten sample fir trees. These ten samples were systematically located within each plot. Amount of defoliation on each new shoot was classified in quarters: ¼, ½, ¾, and complete defoliation. The estimates were weighted *1*, *2*, *3*, and *4*, respectively, and a weighted average computed for each plot. This defoliation index was compared with aerial

---

1. One chain equals 66 feet.

visual and photo estimates of defoliation in other studies. In the study reported here, the index was used to select representative plots in each defoliation-larval population category.

Balsam fir mortality appeared on some plots as early as 1961. Each year thereafter, a 100% cruise was made on plots where heavy defoliation had been recorded for several years. Each dead fir exceeding 4.6 inches in diameter was tallied by species, d.b.h., dominance class and number of sticks of pulpwood in the merchantable height. These data were used by Weber (1964) to develop aerial volume tables for estimating cubic-foot mortality losses in the epidemic area. In addition to tallying dead balsam fir, a count was made of all live fir, white spruce, and black spruce.

Aerial Photography

Large-scale color photography must meet three requirements. First, the camera must have a fast shutter to reduce the image movement at the instant of exposure. Second, a camera recycling rate of up to 5 exposures per second is necessary to obtain a minimum of 60 percent overlap in the flight direction for stereoscopic coverage. Third, color film emulsions must be sensitive so that fast exposure speeds may be used.

We found two cameras that would fit our requirements: the Hulcher 70 Model 102 and a Maurer KB-8 (Heller, Aldrich, and Bailey 1959), (Aldrich 1966).[2] Both cameras, equipped with 6-inch Schneider Xenotar lenses, were used to acquire imagery in this study. Two different color films were used. From 1958 through 1964, Super Anscochrome (ASA 125) was the standard film. In 1965, a change was made to the new Anscochrome D/200 (exposed at ASA 160).

Immediately after we acquired data on the ground, we flew over each plot and took color photographic strips. In 1958 and 1959, we used a scale of 1:1980 (2½ chains to the inch). In 1960, and each year after, the scale was 1:1584 (2 chains to the inch) because we found that more image details were necessary.

Photo Interpretation

Although photography is available for nine years, only four years were used for comparison: 1958, 1962, 1964, and 1966. Since mortality was not detected on the ground until 1961, it seemed of little use in interpreting photography for the years 1958 to 1962 to measure vegetation changes. Because changes occur slowly, two-year intervals seemed adequate to detect indications of change in the vegetation.

The four corners of each plot were accurately located on all sets of photography. Corner markers were referenced to actual one-acre

---

2. Trade names and commercial products or enterprises are mentioned solely for necessary information. No endorsement by the U. S. Department of Agriculture is implied.

plot corners using plot diagrams made on the ground. Once they were marked on one set of photographs, the corners could be transferred to each of the other sets by cross checking. A transparent overlay was made for each plot showing the plot margins in black ink.

Two measures of species distribution and cover type distribution by crown counts are used, the crown count of individuals and crown frequency percent. The latter is the total count of individual crowns in a cover type category, divided by the total in all categories, multiplied by 100. To make the counting task easier, we built a series of transparent templates that divided the one-acre plot into 160 equal squares. These templates were made for scales ranging from 1:1980 up to 1:1056. To count trees by visible crowns, the interpreter selected the appropriate template to fit the scale of the plot. The template was always oriented in the same manner on all photographs. This procedure makes it possible to reorient the template if any further interpretation is necessary.

Each plot was examined with the aid of a 70-mm viewer. All dead and live trees in the overstory were recorded by species and crown dominance class. Twelve tree species were recognized in the study areas:

Poplar (*Populus* sp.)

White birch (*Betula papyrifera* Marsh)

Red maple (*Acer rubrum* L.)

Black ash (*Fraxinus nigra* Marsh)

Jack pine (*Pinus banksiana* Lamb)

Balsam fir (*Abies balsamea* (L.) Mill)

Red pine (*Pinus resinosa* Ait)

White pine (*Pinus strobus* L.)

White cedar (*Thuja occidentalis* L.)

Tamarack (*Larix laricina* (Du Roi) K. Koch.)

Black spruce (*Picea mariana* (Mill.) B.S.P.)

White spruce (*Picea glauca* (Moench) Voss)

The photo characteristics for 10 of these species were described by Heller *et al.* (1964). He found that trained interpreters could identify tree species with a mean accuracy of 90%. Morphological and color features contributed most to a correct identification. Two species found in the present study, but not listed by Heller, were black ash and tamarack. Though not abundant, these two species are easily identified. Both were found on moist sites. Black ash appears bright green on color film. Its compound leaf, arranged opposite each other on the same twig, gives the tips of columnar branches a broad crosslike appearance. Tamarack has an acute-shaped top and a lacy appearance because of its long radiating slender branches. The light blue-green color of its foliage on the film is also a helpful characteristic.

The interpreter used in this study had taken part in the species identification study completed in 1964 and described above. Before starting the interpretation for the present study, tree species characteristics were reviewed and selected examples studied.

Crown dominance on the photos was arbitrarily defined to conform with ground dominance as nearly as possible. This classification is not an exact art—even on the ground—and some variation can be expected. We used the following rules commonly used by silviculturists:

Dominant: generally a taller tree than its neighbors and receiving light from the top and all sides.

Co-dominant: generally of the same height as its neighbors and receiving light on the upper one-third of the crown but comparatively little on the sides.

Intermediate: generally of the same height or shorter than its neighbors and receiving light only on the crown tip.

Suppressed: (this classification was not used because suppressed trees are by definition overtopped and not visible on vertical photography).

The distinction between dead and live balsam fir on photographs is not always clear. Dead fir trees are either white or a pale gray, or they may have a few remaining red needles throughout the crown. Newly dead fir retain their full complement of fine twigs, and the crowns appear finely textured throughout. An older kill appears coarser in texture as it loses its small twigs and branches. Eventually the crown is skeletonized, with only the larger branches remaining. At this stage many dead firs blow down or break off in wind storms. Severely defoliated fir that have little or no new growth, but have some remaining old needles in the lower crown, appear a grayish-green on the photographs. These trees are classified as living trees, though they may die within the same year.

Another photo measure of species and cover type distribution is cover closure. Closure, as used here, is defined as the ground area covered when tree crowns are projected vertically to the ground. Using this definition, we estimated on a dot template cover closure for each plot. The 1958 photography was interpreted with a 200-dot-per-square-inch grid; 1962 through 1966 photography was interpreted with a 144-dot-per-square-inch grid. The change was made to give approximately the same sampling intensity at the two photograph scales—1:1980 and 1:1584, respectively. Despite this compensation, intensity of sampling varied because of the random variations in photographic scale that occur between plots each year.

Each template dot within the plot was examined and classified as it corresponded to the following tree or open ground categories:

Overstory: classified by recognized tree species (12 species); dead and live trees were tallied separately.

Understory: trees less than half the height of the overstory stand.

Open ground: included exposed soil, rock, grass, fern, and herbaceous ground cover.

RESULTS AND DISCUSSION

Spruce budworm larvae feed on new staminate flower buds, vegetative buds and eventually on new expanding growth. Heller and Schmiege

(1962) found that 10 larvae per 15-inch branch sample will consume 75% of the new needles. Five larvae per branch can consume between 25 and 50% of the foliage, depending on how long the population has been at this level. Repeated heavy defoliation, with 50% or more of the new growth removed, can eventually kill a tree. Death can come after three years of heavy defoliation (Bean and Waters 1961), but it usually requires more than three years.

Fir Mortality and Budworm Populations

We examined the relationship between year of defoliation, mortality and spruce budworm populations (Fig. 1). On Plot 12, for instance, budworm populations remained at a level of 10 larvae per 15-inch branch from 1958 through 1964, a period of 7 years. In 1965 and 1966, the populations dropped to less than 6 larvae. No fir mortality estimates were made on the ground before 1962, when 9% of the balsam fir had already died from the prolonged heavy defoliation of current growth. In the 4-year period ending in 1966, fir mortality rose to 52%. The accumulated effect of defoliation will continue to kill trees in these stands.

Photo estimates of fir mortality rose from zero in 1958 to 58% in 1966. In general, they followed the same pattern as mortality determined on the ground but were 5 to 10% higher. The difference is due to trees less than 4.6 inches d.b.h. that were counted on the photographs, but were not recorded in the ground survey.

In 1964, both photo and ground estimates suggested a drop in mortality rate. This decline could mean that there was a slowdown in mortality and some fir trees will eventually survive. As trees are killed and blown down, photo and ground estimates converge because ground count is accumulative from one year to the next as new skills are being added to the total each year. The photo method requires a new count of total fir mortality each year. Blowdown that occurs in the dead fir of previous years is not counted. On areas of severe mortality, the surviving population of trees may prove easier to evaluate and compare.

In another example on Plot 52, budworm populations dropped from a high of 16 larvae per 15-inch branch in 1958 to less than 5 in 1959. From 1959 to 1966, populations dropped to a low of 2 larvae per branch. Thus, when the high insect population dropped, it allowed the fir to recover, and the result was a low mortality rate. On this plot, the fir mortality by photo and ground estimates were within 10% of each other. The photo estimate followed the same pattern as the ground count until 1966, when the photo estimate fell below the ground estimate by 6%. This difference was probably caused by blowdown in the smaller than 4.6-inch d.b.h. group.

Live and Dead Fir Counts

Visible crown counts of dead fir made on large-scale photographs were inconsistent with ground stem counts (Table 1). A number of reasons may account for this difference. Most important is the fact that ground

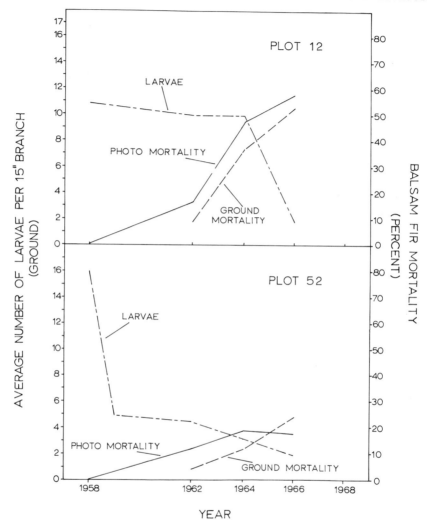

Figure 1. Balsam fir mortality estimated from color photographs compared with a 100% ground cruise of fir mortality and mean larval populations per 15-inch branch sample 1958-1966.

data included only trees 4.6 inches d.b.h. and larger in all dominance classes. Our ground observations did not support the hypothesis that budworms attack only trees larger than 4.6 inches d.b.h. Saplings, 3 to 4.6 inches in d.b.h., were attacked with equal intensity. As a result, the photo interpreter, unable to distinguish between tree diameter classes, tallied many trees in the overstory less than 4.6 inches d.b.h. This fact is reflected in the higher live and dead tree counts on Plot 52.

Between 1964 and 1966, the ground data indicated that many of the dead fir less than 4.6 inches d.b.h. were blown down. For instance, Plot

52 shows that 75 trees counted in 1964 were reduced to 56 trees two years later. This deduction appears logical because smaller trees will have a higher percentage of sapwood, and take less time for blue stain fungus and decay organisms to make them susceptible to breakage. On the other hand, Plot 12 was predominantly a mature stand where fewer trees were susceptible to blowdown during the period of the study.

Since we are concerned with balsam fir survival and changes in stand structure, total live fir counts made on the photographs will show this best. Live fir counts include both mature trees and trees in the sapling size class as well. If live and dead fir counts are compared in terms of

Table 1. Counts of Live and Dead Fir Trees Made on Aerial Color Photographs and 100-percent Ground Stem Counts on Two Plots, 1962, 1964, 1966.

Plot		Ground[1]		Photo[2]	
No.	Year	Live fir	Dead fir	Live fir	Dead fir
[3]12	1962	187	17	181	36
	1964	130	74	101	90
	1966	98	106	68	93
[4]52	1962	284	12	367	49
	1964	262	34	328	75
	1966	240	56	268	56

[1]Includes only trees 4.6 inches d.b.h. and larger.
[2]Includes all trees in overstory.
[3]Severe defoliation and 10 larvae per 15-inch branch.
[4]Moderate to heavy defoliation and fewer than 10 larvae per 15-inch branch sample.

percentage of total fir count, ground and photo counts show a closer relationship (Table 2). In fact, estimates were all within 10%. Again, differences can be attributed to the minimum d.b.h. set for ground data, i.e., more small stems (smaller than 4.6 inches d.b.h.) were counted on the aerial photos than on the ground.

Table 2. Count of Live and Dead Fir Trees Made on Aerial Color Photographs and 100-percent Ground Stem Counts on Two Plots, 1962, 1964, 1966. Photo crown and ground stem counts are expressed as a percent.

Plot	Year	Ground[1]		Photo[2]	
No.	Year	Live fir	Dead fir	Live fir	Dead fir
12[4]	1962	91.7	8.3	83.4	16.6
	1964	64.4	35.6	52.9	47.1
	1966	48.5	51.5	42.2	57.8
52[5]	1962	96.0	4.0	88.2	11.8
	1964	88.5	11.5	81.4	18.6
	1966	81.1	18.9	85.9	14.1

[1]Includes only trees 4.6 inches d.b.h. and larger.
[2]Includes all trees in overstory.
[3]Percent of all fir, dead and alive.
[4]Severe defoliation and 10 larvae per 15-inch branch sample.
[5]Moderate to heavy defoliation and fewer than 10 larvae per 15-inch branch sample.

Tree Species Distribution

We computed the range in percent frequency for the six most prevalent tree species found in three one-acre plots (Table 3). The range includes interpretation at four time periods. Fir was not included because mortality and blowdown would make comparisons invalid. Most of the species were counted consistently on the four sets of photographs.

Among the six species, white spruce showed the widest range. This finding indicates difficulty in interpretation of this species. Since large mature white spruce are very distinct in both hue and morphological characteristics, the problem must lie with smaller immature trees. Birch is another problem species. Unlike white spruce, however, immature trees are easily identified by hue and texture. Crowns of larger trees on the other hand "break up" and look more like old growth aspen. The color difference is not so apparent under these circumstances. Counts of other tree species do not vary more than 4% from one set of photographs to the next. Variations from year to year are small and erratic making it difficult to measure changes in species composition. As time passes, and if one species assumes dominance over the others, changes may be measurable on the photographs.

Table 3. Range in Frequency Percent For Six Tree Species Photographed in 1958, 1962, 1964, and 1966.

Species	Range in frequency percent		
	Plot 12	Plot 52	Plot 112
Poplar	13-15	5-9	5-8
White birch	1-8	8-10	1-2
Red maple	5-10	4-6	0-1
White pine	0	1-4	0
Black spruce	0-4	0-3	2-5
White spruce	3-12	1-3	0-1

When the species are grouped into cover types, better comparisons can be made to show what is happening to the forest community (Fig. 2). For the purpose of this study all hardwoods (angiosperms) were grouped together and all conifers (gymnosperms) other than balsam fir were grouped as "other conifers."

Over the 8-year period of the study, there has been little indication of a change in the hardwood or other conifer components of the three forest communities. The small changes from one set of photographs to the next are erratic and not significant. These irregularities might be caused by photographic scale variations that affect the interpreter's ability to distinguish and separate individual crowns. Variations in plot locations within the stereo model can also affect counts. For example, a plot located along one edge of the photograph may cause crown displacement and obscuration of nearby trees.

Differences in balsam fir counts have been more dramatic, and for this reason a significant trend can be seen on the severely defoliated

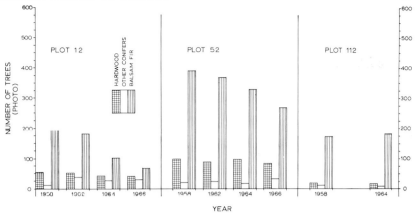

Figure 2. Cover type distribution, by frequency of crowns counted on aerial color photographs for four years.

and moderately to heavily defoliated plots (Plot 12 and Plot 52). Since balsam fir is easily identified, fir counts are acceptable regardless of photo scale.

Live Fir, Understory and Open Ground by Closure Percent

Photo estimates of cover closure for live fir, understory and open ground at the beginning of the study in 1958, and again in 1966, reflect changes in fir cover type (Fig. 3). As might be expected, fir cover type was reduced most on the three plots representing severe defoliation. When defoliation was less, the effect was not so pronounced. Fir cover type increased on plots where defoliation was light and tree growth expanded normally.

In general as fir cover type decreased, understory cover and open ground area increased. As fir cover type increased, understory cover increased, but the amount of open ground area decreased. There were exceptions that could have been caused by sampling errors introduced by using the same dot template on each yearly series of photographs that had some scale variations. And there is an interpreter error. The 1958 photographs were of smaller scale (1:1980) than the succeeding years (1:1584) resulting in openings that were smaller and more difficult to detect and interpret.

Cover Type Distribution

Cover type distribution is difficult to visualize; therefore, we used counts of individual tree crowns by cover converted to frequency percent and cover closure percent to construct phytographs (Fig. 4, 5, and 6). Each of the three defoliation-larval population categories is represented . Phytographs for the initial year (1958) and the year of beginning fir mortality (1962) were compared with phytographs for 1964 and 1966. Photographs beside each phytograph were enlarged to a common scale

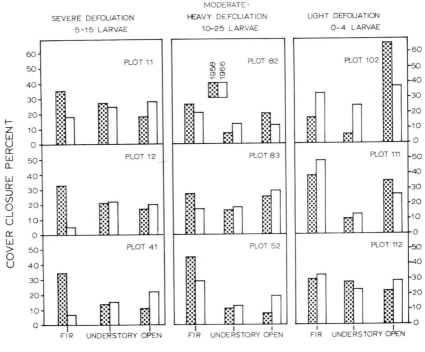

Figure 3. Comparison of cover closure percent for live fir, understory and open ground on nine plots in 1958 and 1966.

from aerial color transparencies. They provide a photographic representation of each plot for ocular comparisons.

The change in both frequency percent and cover closure percent is quite obvious for Plot 12. The fir was reduced in frequency from 75% in 1958 to 49% in 1966. The reduction in cover closure was even greater, from 34 to 6% of the area. This decline means that many large, dominant trees that occupy large areas have died, opening the canopy. In 1966, hardwood overstory had increased from 15 to 23%, other conifers from 13 to 24% and combined understory and open ground from 38 to 47%. It is apparent that the reduction in fir has increased hardwood competition and further changes can be anticipated.

Plot 52, a moderately to heavily defoliated plot with lower budworm populations, showed reductions in overstory fir. However, the change was not as drastic as in the previous example. The reduction in crown frequency was from 77 to 70% and in cover closure from 46 to 30%. Again, it appears that a few larger trees have succumbed to reduce cover closure 16%. Hardwood frequency, though showing a slight increase, did not change appreciably. Hardwood cover closure remained almost the same. The greatest visible change is in "other conifers." How-

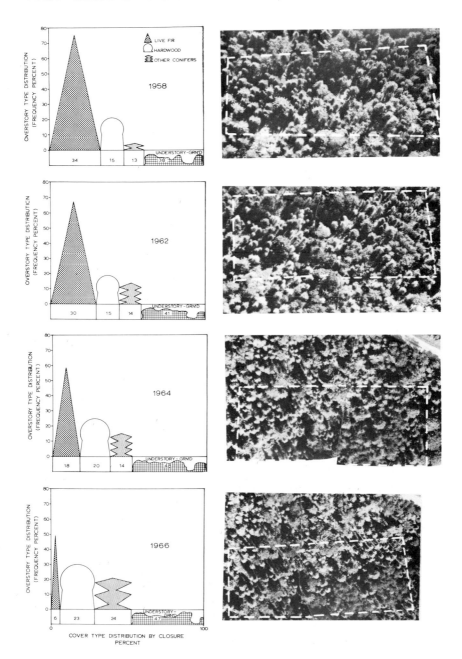

Figure 4. Phytograph for Plot 12 showing cover type distribution, by frequency percent and cover closure percent for 1958, 1962, 1964, and 1966. Photographs were made from color transparencies to show, at a common scale, the changes that have occurred.

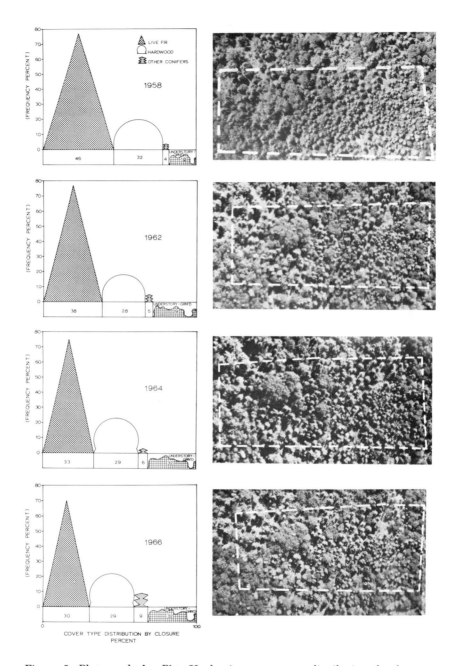

Figure 5. Phytograph for Plot 52 showing cover type distribution, by frequency percent and cover closure percent for 1958, 1962, 1964, and 1966. Photographs were made from color transparencies to show, at a common scale, the changes that have occurred.

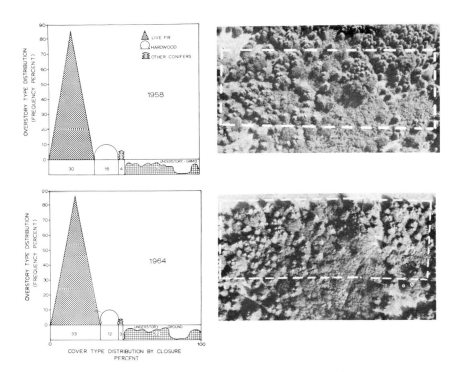

Figure 6. Phytograph for Plot 112 showing cover type distribution, by frequency percent and cover closure percent for 1958 and 1964. Photographs were made from color transparencies to show, at a common scale, the changes that have occurred.

ever, this type had such a minor role in the stands that small changes appear to make big differences in the phytograph. Openings and understory vegetation increased by 14%. As time passes, these new openings should be closed by adjacent trees or understory fir.

The last phytograph (Fig. 6) represents the light defoliation category. Obviously, few changes have occurred. This plot suffered no fir mortality and the frequency percent remained almost constant, but cover closure has increased about 3% through growth. Hardwoods, other conifers, and understory and open areas as well, changed relatively little in the six years.

CONCLUSIONS

This is a continuing study, and the results should not be considered final. But on the basis of the findings to date we can draw several conclusions about the feasibility of large-scale aerial color photography for measuring the changes in a forest community—particularly changes resulting from logging, land clearing, fires, windstorms or insect and disease epidemics:

1. Individual tree crown counts converted to frequency percent are a better measure of change than the counts themselves.

2. Cover closure percent is an excellent measure of the area occupied by overstory and understory components of the forest community. This percent reflects changes in the unoccupied space (aerial and soil) within the community.

3. When frequency percent and cover closure percent are used together, changes over time can be examined in phytographs.

4. Large-scale aerial photograph interpretation will be more reflective of changes in species components of the community if the photographic scales are always the same. This goal might be accomplished by using more precise altimeters during the photography. If this is not practical, photographs can be rescaled by photographic laboratory techniques. Another possibility is to use viewers that will change scale to a constant by optical projection methods.

5. Cover closure percent on differing photographic scales will be more precise if dot templates are scaled to the photograph. Sampling intensity—number of dots (points) examined—will be equal regardless of scale.

6. Balsam fir mortality and surviving live fir can be counted accurately on large-scale photographs. Counts are difficult to relate to ground counts unless the ground count includes all trees in the overstory regardless of size. For precise ecological studies, such detailed counting would be needed.

7. On the basis of a three-plot comparison used in this study, many balsam fir are still surviving. This is true even in stands severely defoliated for several years by high budworm populations. However, because the plots used in this study had relatively high hardwood components, this statement may not hold true in pure fir stands.

SUMMARY

Large-scale aerial photographs were taken annually during a 10-year period over several spruce-fir stands in northeastern Minnesota. The photographs were used primarily to develop aerial photographic sampling techniques for assessing damage caused by spruce budworm (*Choristoneura fumiferana* Clem.). Damage to spruce-fir stands was assessed on the photographs and related to budworm populations and damage measured on the ground. A secondary purpose was to follow the history of mature spruce-fir stands during and following an epidemic. The color photography provided a good assessment of changes in balsam fir density and permitted identification of boreal tree species with high accuracy. Phytographs based on photo images showed differences in species frequency, composition, and crown cover. The use of large-scale color photography appears to be a useful tool in assessing forest community changes.

REFERENCES

Aldrich, R. C. 1966. Forestry applications of 70 mm. color. Photogr. Engr. 32:802-810.

Bean, J. L. and W. E. Waters. 1961. Spruce budworm in eastern United States. U. S. Dept. Agr., Forest Pest Leaflet 58:8.

Heller, R. C., R. C. Aldrich, and W. F. Bailey. 1959. Evaluation of several camera systems for sampling forest insect damage at low altitude. Photogr. Engr. 25:137-144.

Heller, R. C., G. E. Doverspike, and R. C. Aldrich. 1964. Identification of tree species on large-scale panchromatic and color aerial photographs. U. S. Dept. Agr., Agr. Handbook No. 261:17.

Heller, R. C. and D. C. Schmiege. 1962. Aerial survey techniques for the spruce budworm in the Lake States. J. Forestry 60:525-532.

Ryan, S. O. and H. O. Batzer. 1964. Spruce budworm defoliation in northeastern Minnesota decreases in 1963. U. S. For. Serv., Lake States For. Exp. Sta., Research Note LS-39:2.

Weber, F. P. 1964. An aerial survey of spruce and fir volume killed by the spruce budworm in northern Minnesota. U. S. For. Serv., Research Note WO-2:5.

Ecological Applications of Multispectral Color Aerial Photography

EDWARD YOST and SONDRA WENDEROTH

THE EXPERIMENT AND THE OBJECTIVE

REMOTE sensing is a valuable tool for the ecologist. It permits him to extend observation of the relationships between living organisms and the environment to vast areas otherwise impossible to investigate. Remote sensing is important to the scientist in allowing him to determine the spatial relationships of the environment by synoptic view of interaction of ecological variables such as soil types and soil moisture with the geographical distribution of plant species.

A remote sensing experiment using a Multispectral Color Photographic System was conducted in the summer of 1967 over agricultural and forestry test sites of the National Aeronautics and Space Administration (NASA) located in California. One of the purposes was to establish the utility of this type of remote sensing system for measurement of basic ecological parameters and to compare the results with conventional color and infrared color film. A secondary objective was to explore possibilities of obtaining unique signatures for species of agricultural crops, trees and soil types based upon this photographic method of combined spectroradiometry and colorimetry.

The Multispectral Color Photographic System (Yost 1967) combines abridged spectral photographic data collection in four spectral bands in the 360 to 900 mμ region of the spectrum with colorimetric data reduction using additive color techniques. This remote sensing technique (Yost and Wenderoth 1966, 1967, 1968) allows the scientist to collect spectral photography in the near-ultraviolet, visible or near-infrared region of the spectrum and provides him with a device to combine the photos into a single color picture in a viewer. This viewer also provides the scientist with the capability of altering the color of the presentation in order to enhance the particular relationships.

There are three classes of problems associated with remote sensing of the environment which should be examined at the outset. The existance of unique spectral signatures for most environmental parameters of ecological interest are not known. There is an almost complete absence of reliable spectral data taken *in vivo*. Dynamic variables in the environment such as variation in illumination, atmospheric attenuation and preferential (non-lambertian) spectral reflectance of the natural objects exist which distort the underlying relationships being sought. Secondly, instrumentation errors such as variation in spectral illumination in the focal plane of a camera sensor and the dependence of photographic density on wavelength of radiation can result in errors in photographic reproduction. Finally and perhaps the most significant problem which is often overlooked, is the physical fact that a unique spectra produces a unique color, but that the inverse is not true. The same color may be made by an infinite number of different reflectance spectra. The technology of multispectral color photography is designed to circumvent this physical law of nature.

In order to quantitatively examine these environmental and instrumentation variables, man-made targets of known spectral characteristics were placed in the environment. In addition, measurements of the incident solar radiation, spectral reflectance and color of the standard targets were made at the time the spectral photography was obtained. Laboratory calibration of the multispectral camera, control of the photo reproduction process as well as analysis of the viewer color variables were performed to reduce instrumentation errors to a minimum.

The objective of the experiment was then to combine abridged spectroradiometry and colorimetry so that a unique color could in fact be presented to the interpreter which would be representative of a particular ecological parameter (e.g., tree specie) reducing environmental and instrumentation variables to an absolute minimum.

THE INSTRUMENTATION

The multispectral camera and viewer were designed as an integrated system. The camera takes four spatially identical photos, the images being recorded on the film in precise location with respect to a common coordinate system. Filtration into four different spectral bands causes density differences in these four images of a ground object when there exists a variation in the spectral reflectance of the object. The viewer optically superimposes the four photographs and presents the interpreter with a single presentation in which the density differences between the four spectral photos appear as colors. The color space can be manipulated by the interpreter.

The four lens multispectral camera, shown in Figure 1, takes spectral photos in any four bands from 360 to 900 mμ which covers part of the ultra-violet, the visible and part of the near-infrared spectrum. The camera is designed to take a set of four spectral negatives at the same time and to record them on one piece of film. Because each band is photographed through its own lens, it is possible to obtain the correct exposure when large differences exist in the spectral radiance of any scene. This control of exposure for all four bands permits repeatable accuracy under a wide range of illumination and ground spectral reflectance conditions.

Each one of the four spectral records (Fig. 2), which together comprise a set of multispectral photographs, is taken at exactly the same time by the four camera lenses. Since the optical axes of all the lenses are normal to the film plane, and all are color corrected to give the same focal length and distortion throughout the entire spectral region

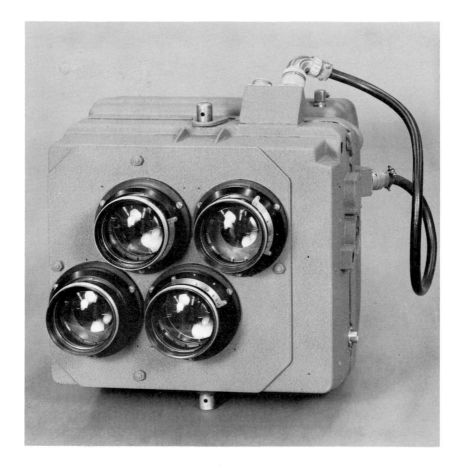

Figure 1. Multispectral Camera Used in Research.
    The camera contains four 7 inch focal length f/2.5 lenses which are precisely matched in both focal length and distortion. The magazine allows either a single piece of film for all four images or four individual pieces of film. It incorporates image motion compensation.

(360 to 900 mμ), four spatially identical photos are produced. All images appear in identical coordinate positions as measured from the principal point of each photograph.

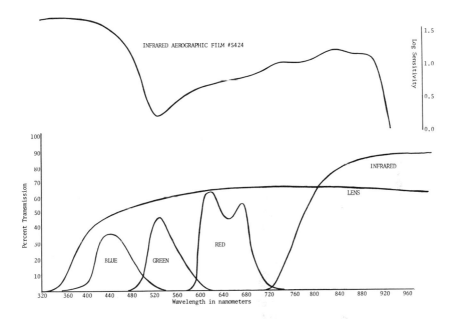

Figure 2. Photo-Optical Camera Characteristics.
Lower curves show percent transmission of the four filters used in the experiment and the lens transmission on axis. The upper curve shows the log sensitivity of the infrared sensitive film indicating reduced exposure index (speed) in the green region.

The sets of four spectral negatives must be developed so that the density of the image on each individual negative is a correct representation of the brightness of the object. In addition to compensation for differences in exposure, it is necessary to correct for reduced contrast which may exist in one of the negatives due to the wavelength of the radiation, in order to produce an identical relationship of exposure to density on all four negatives (Fig. 3). A gamma (the slope of the exposure-density curve) is chosen to produce good contrast (slope between one and two) without excessively reducing the exposure range.

The rear projection viewer (Fig. 4) is used to form a composite color rendition of these spectral positives. The optical system of the viewer allows control of the color sensations of brightness, hue and saturation. Illuminating each spectral positive by a different color of controlled brightness and chromaticity and at the same time superimposing each spectral positive on the screen, one upon the other, produces a composite color presentation (Yost and Wenderoth 1965).

The color of the image presented on the viewer screen, frequently

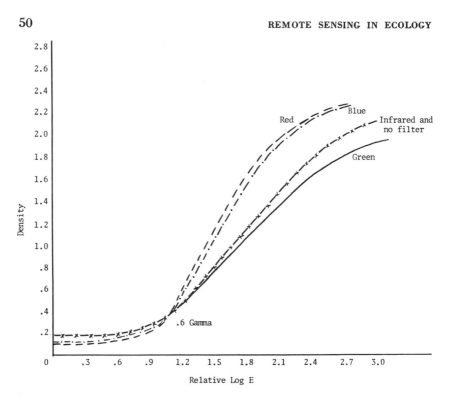

Figure 3. The characteristic curves showing exposure- density relationships for infrared film (#5424) when the filters shown in figure 2 were used.

called its position in the color solid, can be manipulated by variations of the brightness lamp containing a filter in the optical path and a saturation lamp. The light from the brightness lamp controls the brightness of the illumination of each spectral positive. The filter establishes the hue and the illumination of the saturation lamp controls the amount of desaturation of the color. Manipulation of these color variables (Fig. 5) by the scientist permits the detection of subtle differences in the images of ecological variables in the environment.

RESULTS OF THE EXPERIMENT

The multispectral system is designed for scientific analysis of remotely sensed spectral photography by viewing the composite additive color image projected upon the viewer screen. However, in order to present some results of interest to ecologists, reproductions of significant multispectral color screen presentations by subtractive color copying process have been made. Considerable care has been taken to make the color fidelity of the reproduced image as close as possible to that observed on the viewer screen, but of course perfect reproduction can not be achieved in color printing.

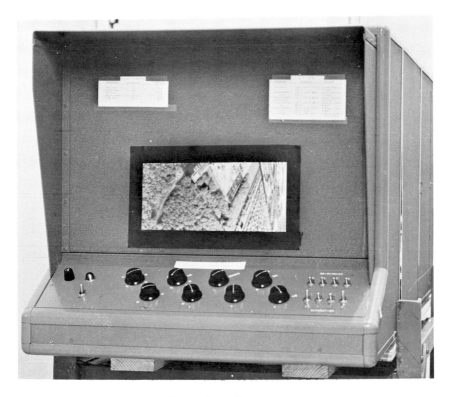

Figure 4. A Multispectral Additive Color Viewer.
This interpretation apparatus optically superimposes four black and white
spectral positives on a rear projection screen and shows all density differences
between the spectral positives as color. The controls on the front of the viewer
permit the interpretor to adjust the hue, brightness or saturation of the color
image.

Four simultaneous positive photographs taken by the multispectral
camera in the blue band (395 to 510 mμ), the green band (485 to 590 mμ),
the red band (585 to 715 mμ), and the infrared band (700 to 900 mμ)
are shown in Figure 6. The photography was taken over the Agricultural
Experimental Test Station, Davis, California at an altitude of 28,000
feet above sea level at 1200 Pacific Daylight Time on 31 July 1967.
The scale of the photography is approximately 1:44,000. Many density
differences can be detected within each multispectral record as well
as between the four records. However, attempts to identify signatures
of ground objects by multiple comparison of density differences on
the four photographs is difficult and tedious at best.

By registered superimposition of the four records in the multispectral
viewer, using the color space control, a composite color image is presented in which all density differences between records appear as a color.
Where no density difference exists the image is a shade of grey. Using

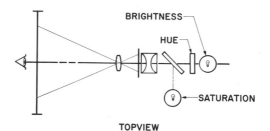

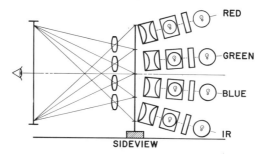

Figure 5.

    This schematic of the viewing system shows how the position of an image
in the color solid is controlled by the viewer illumination system. The top view
shows this arrangement for one spectral position. The luminous output of the
brightness lamp passes through a filter, a condensing lens and illuminates
the spectral positive which is projected on the screen at the left. The amount
of desaturation of the color is varied by changing the luminous output of the
saturation lamp. The bottom view shows repetition of this illumination system
for each spectral position and how one piece of film containing four spectral
photos may be optically superimposed.

this method, the two hundred perceiveable density differences (Fig. 6)
are expanded into 7,500,000 perceiveable color differences. Three of the
many possible multispectral color renditions resulting from the projection of the photography are shown in the color plates.

    An interesting application of this remote sensing technology to
determination of crop species and yield analysis is shown in Figures 7
through 10. The reproductions shown are multispectral color photographs
produced by the additive color viewer from the four black and white
spectral positives shown in Figure 6. Significant ground truth is presented on the annotations shown in Figure 6 and has been keyed to the
following discussion.

    The large corn field (C) can be seen (Figures 9 and 10) to vary in
color. This color change was related to the relative maturation of the
crop which was five to six feet high at the top and just sprouting at
the bottom of the photo. This effect was difficult to detect in the
visible color renditions (Figures 7 and 8) although the demarcation
of soil and foliage can be detected. The visible color renditions were
useful in differentiating plowed fields (P) from stubble (S).

Variations in the development of tomatoes is shown by analysis of the areas (E) on Figures 7, 9, and 10. E_1 is a mature field of tomatoes in which variation in plant density indicates fertility differences in soil. E_2 is a plot planted later than E_1 in which the plants have completed flowering. The discoloration visible in Figure 10 results from standing irrigation water. The tomatoes in field E_3 are similar to E_2 but have not been irrigated for four days. E_4 is a recently planted field in which the plants have not yet flowered.

Plots marked N show the spectral differentiation of beans in which N_1 were mature plants and N_2 recently sprouting plants from 0.5-1.0 ft. tall.

Alfalfa (A) was readily detected in both Figures 8 and 9 as a homogeneous bright red color. However, in visible color (Figures 7 and 8) alfalfa appears quite similar to both corn and tomatoes. Beans and tomatoes grown in similar soil conditions and irrigated simultaneously in adjacent plots can be readily differentiated (E, N, E) in both infrared color renditions at the bottom of the page. The characteristically bright red appearance of plot (M) in Figure 9 and 10 indicates a patch of mature melons. The pasture (PA) can be identified by a mottled appearance on infrared color photograph. Pasture is frequently misidentified as a field of tomatoes in visible color photographs. Almonds (AL) and walnuts (W) are two groves which can be differentiated in color in both Figure 9 and 10 although they were quite similar in color in Figure 8.

A significant example of the distribution of tree species is indicated in Figure 12 which is a multispectral color photograph of the area adjacent to Silver Lake, California. The photography was taken at an altitude of 28,000 feet above sea level at 1330 Pacific Daylight Time on 7 August 1967. The average scale is approximately 1:38,000. The ground truth data has been obtained and is presented in Figure 11. The following discussion is keyed to the annotations contained in Figure 11 and can be referenced to the multispectral color photograph.

The distribution of four species of trees was readily identified by color difference on the multispectral photograph. California Black Oak (*Quercus Kelloggii*) is indicated as bright orange (A); riparian hardwoods, principally White Alder appear as an orange red (D); Lodgepole Pine (*Pinus contorta*) is a dark magenta (G); White Fir (*Abies concolor*) appears as a dark red (H).

The spatial distribution of sedges and grasses in a dry meadow is also clearly color differentiated (C). Variations in shades of pink was related to the vigor of the plants; cyan color indicated bare soil. The concentration of Manzanita cover in open fields can be directly contrasted to the sparseness of the same type of cover in the morains (B). The distribution of White Alder and Manzanita is readily identified as pink and red colors among the granitic outcrops (E) which appear blue.

Figure 6. Positive reproductions of four simultaneous records taken by the multispectral camera. From top to bottom, the blue band records from 395 to 510 nanometers, the green band from 485 to 590 nanometers, the red band from 585 to 715 nanometers, and the infrared band from 700 to 900 nanometers. KEY: C, corn; P, plowed field; S, stubble field; E_1, mature tomatoes; E_2, tomatoes reaching maturity, recently irrigated; E_3, tomatoes just before flowering; E_4, tomatoes before flowering; N_1, mature beans; N_2, beans, recently sprouted; A, alfalfa; PA, pasture; W, walnuts; Al, almonds; M, melons.

Figure 7. A multispectral true color rendition copied from the viewer screen in which the blue, green and red multispectral positives are projected each in the same color.

Figure 8. An aerial ektachrome photo taken at the same time as figure 7 using an optical system which is identical to that of the multispectral camera.

Figure 9. An aerial ektachrome infrared photograph of some agricultural crops exposed at 1617 PDT on 31 July 1967.

Figure 10. A multispectral rendition in which the green band is projected as blue, the red band projected as green, and the infrared band projected as red.

Figure 14 is a multispectral photograph of a serpentine area adjacent to Bucks Lake, California, which was taken at 1300 Pacific Daylight Time on August 7, 1967. The scale of the photography is approximately 1:38,000. Figure 13 is an annotation of the photographs which show data taken at different points on the ground.

Dense riparian hardwoods predominantly Willow (*Salix sp.*) and meadows of dry grass with scattered Willow can be identified (*B*). The distribution of willows has been correlated with the greater moisture in the stand and less soil moisture in the meadow. The moisture of the meadows (*D*) and (*E*) can be inferred by the density of cover of verdant meadow grasses. Meadow *E* contains greater soil moisture than *D* and both meadows, *D* and *E*, have greater average soil moisture than *C*. The existence of lush grasses and sedges (*Carex sp.*) can be identified along with standing water in the wet meadow *G*. Bare soil from gold dredge trailings is shown (*I*) and the color can be related to progesssive change in soil image color tones.

An example of open range land is shown in Figures 14 and 15. This area is Harvey Valley, California (Lat. 40° 41′ N, Long. 121° 04′ W), and was photographed at 1430 Pacific Daylight Time on August 7, 1967. The distribution of meadow grasses that are particularly important for cattle range was inferred from the color variations in the multispectral presentation. Dense stands of sedges in standing water can be seen by the color differences. The state of vigor of rushes and sedges (*Carex sp.*) can be identified with receding water after winter (*1 and 2*). Low sage brush growing on light-toned soil is related to the light brown color (*3*). This type of community was differentiated from a reseeded area of *Bromus inermus* grass (*4*) which had a darker color. The potential productivity of the two areas of grazing forage is greater for area *4* than *3*.

The presence of dense, lush meadow vegetation predominantly sedges, forbs and some water-loving grasses (*5*) indicates high productivity of this forage area. Big sagebrush sites (*6*) can be identified by change in brightness as compared to Low Sagebrush types of cover (*3*). Big Sagebrush covers about 20% of the ground surface and grows in reddish brown soils up to three feet deep. Stands of Ponderosa Pine are readily identified (*7*) as are the Manzanita brushfield which have invaded the forested area following a burn (*8*).

ENVIRONMENTAL MEASUREMENTS

The amplitude and wavelength of electromagnetic radiation in the 360 to 1300 mμ band which exists in the natural environment constantly varies. The intensity and spectral distribution of solar radiation falling upon the ground changes with both solar angle and atmospheric conditions. Objects on the ground reflect varying amounts of this radiation in every wavelength not only in proportion to that falling upon them but also as a function of environmental variables which dynamically change their absorption, transmission and reflection characteristics.

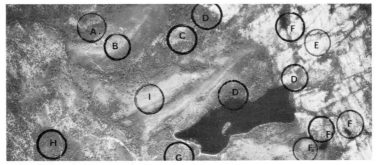

Figure 11. Silver Lake
A-Quercus kelloggii, B-Glacial Moraine (Granitic) with sparse covering of Manzanita brush, C-Dry Meadow of sedges and grasses, D-Riparian hardwoods (white alder), E-Granitic outcrop, G-Pinus contorta, H-Abies concolor.

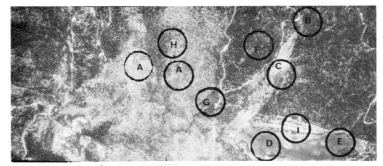

Figure 13. Serpentine Area
A-Dubakella soil. Ground cover is Ceanothus cuneatus (buckbrush), B-Riparian hardwood vegetation predominantly salix sp. (Willow), C-Meadow of dry grass with some scattered salix sp., D-Meadow with a higher proportion of green vegetation, E-Very moist meadow area, F-Dense stand of mixed conifers on cohasset soil, G-Wet meadow, lush grasses and sedges (Carex sp.), H-Fairly dense stand of timber on dubakella soil, I-Bare gravel—gold dredge tailings from early mining operations

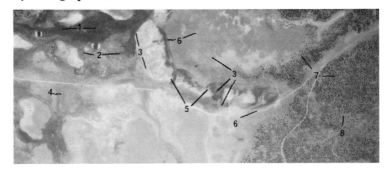

Figure 15. Harvey Valley Range
1-Integration of standing water in and among a dense stand of sedges (Carex sp.), 2-Partially dried rushes and sedges, 3-Low sagebrush type, 4-Bromus inermus grass, 5-Very dense, lush meadow vegetation dominant species include sedges, forbs and some water-loving grasses, 6-Big sagebrush sites, 7-Ponderosa pine, 8-Manzanita brushfield

Figure 12. Silver Lake, California 28,000 feet above sea-level false color rendition. Green band projected as blue, red band projected as green, infrared band projected as red.

Figure 14. Serpentine area of Bucks Lake, California 28,000 feet above sea-level. The green band projected as blue, red band projected as red, the infrared band projected green.

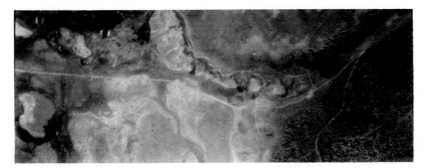

Figure 16. Harvey Valley, California 28,000 feet above sea-level false color rendition. Green band projected as green, red band projected as red, and the infrared band projected as blue.

Figure 17 is a graph of the spectral distribution of solar radiation taken at Davis, California on 31 July 1967 using a spectroradiometer with a lambertian (cosine) detector surface. The large variation in microwatts/cm²/mµ of total sunlight in diffuse skylight falling upon the ground at different times during a clear day is evident. Also this graph shows that at this particular place and time there was a large increase in incident infrared radiation at 750 mµ in the afternoon compared to the morning. This is probably due to the absorption of visable light by large dust particles which had been churned up in the fields during the day. The ability to control the exposure independently in each spectral band gives the multispectral camera the ability to compensate for this dynamic variable in the environment. Color infrared film on the other hand appeared to be more red in the afternoon compared to the morning because of the fixed sensitivity of the individual dye layers in each spectral band.

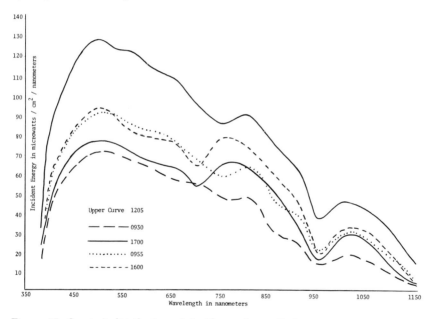

Figure 17. Spectral distribution of incident solar radiation
    The intensity and spectral distribution of solar radiation falling on flat ground
    varies with time of day (solar altitude) and with atmospheric conditions.

The spectral difference between the solar radiation falling directly on the ground and the diffuse skylight which illuminates shadowed areas is one aspect of solar illumination that can not be compensated by variation in exposure in the multispectral camera. Figure 18 shows a comparison of direct sunlight and diffuse skylight made at the same time on 30 July 1967 at Davis, California. In addition to the obvious reduction in intensity, skylight is predominantly blue which results in

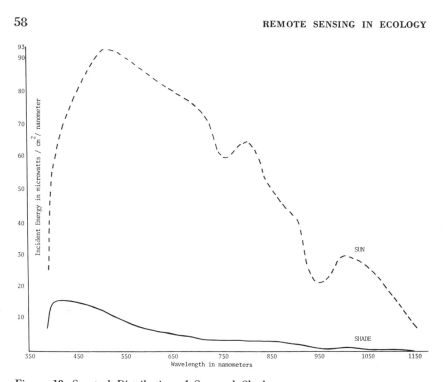

Figure 18. Spectral Distribution of Sun and Shade
Not only is the spectral intensity of sunlight and diffuse skylight (which illuminates shaded areas) different, but also their distribution is quite dissimilar.
Shaded areas are illuminated by light having a larger proportion of blue
resulting from diffuse skylight illumination.

an apparent increase in the relative amount of radiation reflected in
the blue band compared to other bands when an object is in the shade.

Any remote sensing instrument is a perspective sensor. It views an
object on the ground at a specific angle which depends on both the
orientation of the object and the orientation of the sensor. Each image
on a photograph thus is seen as being in a different perspective. Solar
radiation also falls on surface objects at varying angles of incidence
depending on both the time of day and the orientation of the object. The
question then arises, what effect do these angular relationships have on
the spectra of an object as recorded by the sensor?

Figure 19 demonstrates this important and unresolved problem. A
grey target was specially constructed to present a diffuse surface which
was measured in the laboratory using a spectrophotometer with a total
diffusing sphere. In the field, however, a spectroradiometer was employed to measure the incident noon sun and skylight falling onto the
grey panel. The percentage noon sunlight reflected by this "diffuse"
grey panel, which was designed to be used in order to calibrate the
spectral and color photographs, is quite different from the total diffuse
reflectance calibration made in the laboratory. Also, readings were

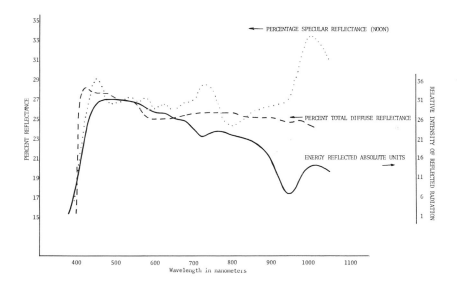

Figure 19. Comparison of the spectrophotometric total diffuse reflectance and percentage reflectance under natural illumination in the direction of the energy actually reflected.

taken in the field using a spectroradiometer which was oriented at a 30° angle with respect to the surface normal and 90° in azimuth from the direction of the sun. The actual amount of solar radiation reflected by the target in the direction of this orientation, which would be the energy recorded by a remote sensor in the same orientation and position Fig. 19, is shown in the solid curve of Figure 20. This curve shows the target spectra which results when the combined effect of not only the spectral distribution and intensity of sunlight is considered but also the directional reflectance characteristics of the target. The distribution of reflectance spectra also depends upon the relative direction of the incident radiation, the perspective angle of the sensor in addition to the orientation of the object.

Chromaticity of four color panels was measured on the ground at the time of flight (Fig. 20). In addition, the colors of the images of the panels as they appeared on the viewer screen and on Ektachrome color film were also determined. Note how closely the additive color image can be made matched to the actual color of the object. The large difference which exists in the color film is due to the inherent problems in color reproduction using dyes. Although the chromaticity coordinates appear close to the white point in the center of the diagram, it should be emphasized that the brightness levels of the panels were only between 21% and 42%.

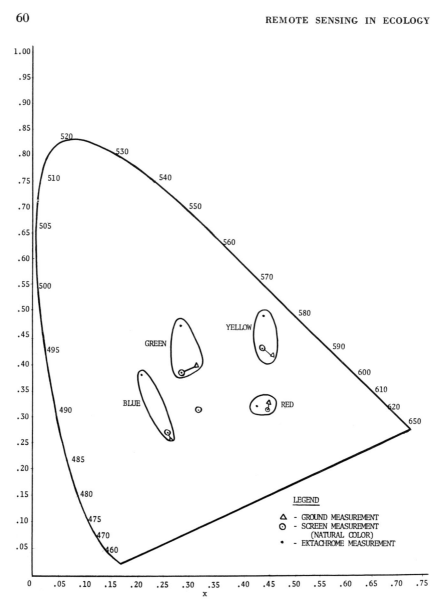

Figure 20. Comparative chromaticity coordinates of color target panels and the multiband color as well as ektachrome color of their images.

Figure 21 presents the spectra of alfalfa, rye and safflower under noon sun conditions on 29 July 1967 at Davis, California. Although the results were dependent on the specific perspective angles which existed at the time of measurement, they do show unique spectral signatures for these crops in the near infrared photographic spectrum.

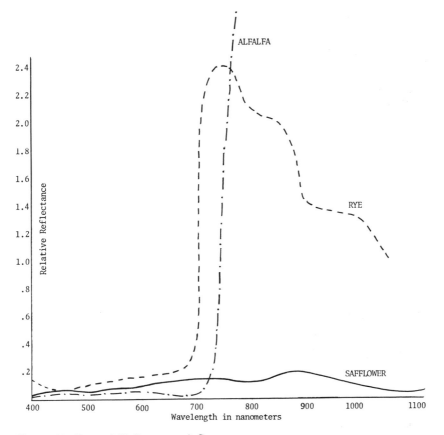

Figure 21. Spectral Reflectance of Crops
  Spectral reflectance of alfalfa, rye and safflower measured in the field under
noon sunlight illumination.

SUMMARY

The summer 1967 remote sensing experiment demonstrated the utility of multispectral color aerial photography to the environmental sciences. The ability to detect and identify basic ecological parameters such as tree species, crop types, crop maturation, soil types and range vegetation using broad-band spectral filtration, which were not *a priori* designed for the task, is most significant. The possibility of eliminating instrumentation errors which have heretofore plagued the technology has been achieved. A reduction of the effects of environmental variables particularly the variation in solar radiation during the day has been accomplished. A beginning toward understanding the cause of spectral errors resulting from perspective has also been made.

The data collected indicate that the near infrared spectrum apparently offers significant possibilities for detection of differences among vegetative species. Particularly, sub-division of the spectral band ranging from 680 to 900 mμ seems to offer promise for establishing unique chromatic differences in the images of living vegetation. The ability to make a color image closely match the actual colors of objects as they exist in the ground environment, correcting the distortion by variations in solar illumination and atmospheric attenuation, is promising for establishing soil types, perhaps because soils have been so often categorized by their apparent color.

ACKNOWLEDGEMENT

The ground truth and environmental measurements were conducted by William Press of Long Island University in conjunction with the following students from the School of Forestry, University of California: William Draeger, Eric Janes, Jerry Lent, Larry Pettinger, John Thomas and Gene Thorley. Robert Anderson at Long Island University made the photographic reproductions.

REFERENCES

Yost, E. and S. Wenderoth. 1968. Additive color aerial photography. *In* Manual of Color Aerial Photography, J. Smith, editor, American Society of Photogrammetry.

Yost, E. and S. Wenderoth. 1967. Multispectral color aerial photography. Photogr. Engr. 33:1020-1033.

Yost, E. 1967. Application of a multispectral color photographic systems. Proc. of the C.I.S./I.C.A.S. Symp. Air Photo Interpretation, Ottawa, Canada.

Wenderoth, S. and E. Yost. 1966. A rapid access color reconnaissance systems. Symp. Human Engineering in Photo Data Reduction, Society of Photographic Instrumentation Engineers.

Yost, E. and S. Wenderoth. 1965. The chromatic characteristics of additive color aerial photography. Photographic Science and Engineering 9:3.

A Multi-Sensor Study of Plant Communities at Horsefly Mountain, Oregon

Rex M. Peterson
G. Ross Cochrane
Stanley A. Morain
David S. Simonett

INTRODUCTION

THE Horsefly Mountain region is a complex of upland pine and fir forests and of near-flat, stony basins supporting grass or sage. Located in south central Oregon, 70 kilometers east of Klamath Falls, it is transitional between the mesic forests of the Cascades to the west and the semiarid grass and sage of the Columbia Plateau to the east. (Fenneman 1931; Atwood 1940; Hunt 1967). A wide range of vegetation communities is present (Figure 1). Thus, it is an excellent area for vegetation analysis, providing distinct boundaries between communities of very different lifeform including grass, low sage shrub, taller chaparral shrub, low tree woodlands and tall tree forests. In addition, important differences within communities, especially variations in height, layering and density partly induced by man's activities—logging, firing and grazing—provide further variety. For these reasons it has been the site of a number of previous studies by several of us (Morain and Simonett 1966; Morain 1967; Morain and Simonett 1967). The present study continues this series, and others are in progress.

More than half the area is covered with ponderosa pine forests of very mixed age, height, structure and density. White fir is common but more localized than pine. Incense cedar and scattered pockets of hard-

This study was supported jointly by the Geography Branch of the U. S. Geological Survey under contract 14-08-0001-10848, for the National Aeronautics and Space Administration, and by the Engineer Topographic Laboratories U. S. Army under Contract DAAK02-68-C-0089 (Project THEMIS).

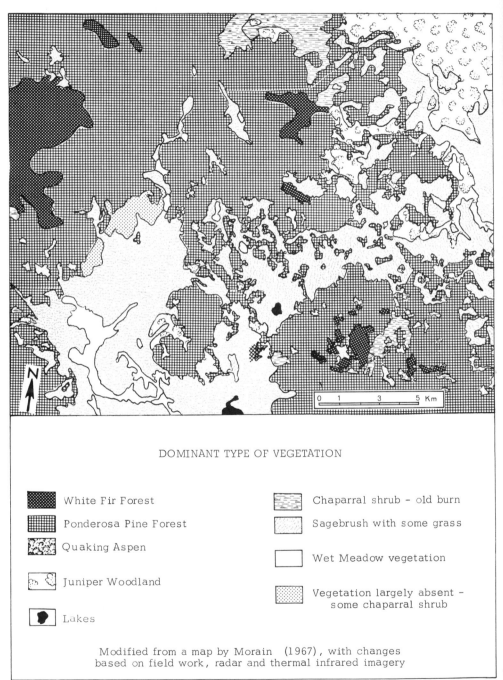

DOMINANT TYPE OF VEGETATION

White Fir Forest

Ponderosa Pine Forest

Quaking Aspen

Juniper Woodland

Lakes

Chaparral shrub - old burn

Sagebrush with some grass

Wet Meadow vegetation

Vegetation largely absent -
some chaparral shrub

Modified from a map by Morain   (1967) , with changes
based on field work, radar and thermal infrared imagery

Figure 1.

wood (quaking aspen) also occur along with juniper woodlands, sage-brush, chaparral on old burns, and various grassland communities. Almost all the plant communities show a notable variation in plant density. Responses to overgrazing or to overgrazing and regrowth following fires and logging are also variable.

In the treatment which follows we discuss in turn: (1) the physical background, noting the diverse lithologies, soils, and slope and aspect characteristics of the region, which contribute to the diversity of the vegetation, (2) the vegetation communities with particular attention being given to life form, plant densities, and boundaries, (3) an electronic image discrimination, enhancement, combination and sampling system (IDECS), developed at the Center for Research in Engineering Science, University of Kansas. This system is used to produce color combined images using multiband, color, radar, and other images. A variety of techniques were used with the IDECS including differentiation, level selection, combination and texture discrimination, (4) finally, a number of color combinations produced with the IDECS are evaluated for their use in delineating vegetation boundaries and in detecting variations in plant density within communities. (Density, as used in this paper, refers to foliage concentration.)

THE PHYSICAL ENVIRONMENT

Wide variations in habitat conditions occur in the relatively small area studied near Horsefly Mountain. Elevations range from 1300 to 2200 meters, slopes from flat to 15 degrees, and rainfalls from 35 cm on the lowlands to around 100 cm on the mountains. In addition, lithologies range from acid to basic, weathering states from weak to strong, and soils from coarse sand to tight clay, from acid to alkaline, and from skeletal to very deep.

Forested, coarsely dissected ridges (Bly Ridge, Gerber Rim) and peaks (Horsefly Mountain, Yainax, Paddock, Tub and James Buttes) of largely basaltic or intermediate volcanic rocks rise steeply from stony lava plains (Plate II, Figure 6) or from shallow alluvial basins (Plate I, Figure 1 and Plate II, Figure 2). Relative relief over much of the area is around 350 meters. With the notable exceptions of Yainax Butte (2203 m) and Horsefly Mountain (1971 m), summit elevations are commonly from 1650-1700 m.

Soils vary considerably throughout the region providing many different habitats for vegetation. Differences in depth, texture, and water holding ability are often closely related to age of parent material. Some of the most recent (late Tertiary) lava flows are still extremely rocky and but little weathered (Plate II, Figure 6). Small pockets of soil between the lava blocks support very sparse, low, xerophytic, sagebrush. Other flows that are more strongly weathered have deeper soils.

A wide mixture of soils alternate closely on the prairies or flats with but little change in relief (Plate I, Figure 1; Plate II, Figure 2; and

Plate II, Figure 2). These include shallow and deep alluvials, residual silts, loams, stony flats with shallow soils, heavy cracking clays, and permanently wet clays in depressions.

There is an abrupt edaphic boundary between the lava plains and the slopes, hills and mountain slopes. This is reflected dramatically in the change from grass or sagebrush communities on the generally young, shallow soils of the flats to the tree communities on the older, deeper soils of the flanking areas. Within the latter, soils vary from stony outcrops, through thin skeletal soils to deeper weathered podzolic soils. Deep krasnozem-like soils are present on some higher southeastern slopes. Partially stabilized areas of wind-blown sand are also present on gentle slopes near Gooch Meadow. There is considerable evidence on Dry Prairie of recent active erosion by sheetwash and wind, which probably resulted from overgrazing. Soil deterioration has caused modification of plant communities. Not only have plant densities been affected but community composition also has altered markedly even to change in life form. Bunch or tussock grass communities have been replaced by less dense, low, shrub communities following sheetwash, wind-stripping and deflation.

VEGETATION

The wide range of vegetation communities present in the Horsefly Mountain region reflects the considerable habitat diversity noted above. Differences in parent material, soil, total and available moisture, aspect, elevation and effects of orographic precipitation influence the distribution and density of the plant communities. Thus, although ponderosa pine (*Pinus ponderosa*) covers more than half of the survey area it is replaced at the wetter limits of its distribution by white fir (*Abies concolor*), at its colder limits by lodgepole pine (*P. contorta*) and at its drier margins by western juniper (*Juniperus occidentalis*). Where soils are shallow with poor moisture retaining capacity sage or grass communities replace pine. It is also replaced by sedge meadows on shallow saturated soils and by quaking aspen (*Populus tremuloides*) on deeper saturated soils.

Effects of fire, of logging, and of grazing exert a secondary influence, adding further complexity to the vegetation mosaic. Chaparral scrub is a normal stage in fire succession in the ponderosa pine communities. Logging practices frequently result in early colonization of disturbed ground surface by species of rabbitbrush (*Chrysothamnus*). Heavy selective browsing by mule deer markedly modifies both the composition and structure of understory shrubs, particularly bitterbrush (*Purshia tridentata*) and mountain mahogany (*Cercocarpus* spp.). Overgrazing of prairie grasses, as noted earlier, causes marked reduction of tuft grasses and an increase in sagebrush distribution. Heavy grazing of sagebrush also results in reduced density, soil deterioration and increasingly depauperate vegetation cover.

Because of this close interrelationship between plant community and its habitat, any measure of plant boundary or plant density carries considerable potential information about the environment. This may have important applications in land utilization planning, forestry, range management, and engineering.

Tree Communities

Ponderosa Pine—Except for the sage and grass communities on the dry, stony or alluvial basins, ponderosa pine is the climax vegetation over most of the area. In undisturbed state it is characterized by relatively dense, even-age stands of simple structure. However, the Fremont National Forest has been extensively logged so pine communities are usually unevenage stands. They are commonly two-tiered or multi-tiered with wide differences in density of layers (Plate I, Figures 2 and 3). Crown canopy coverage differs markedly, especially between scattered, tall (45 m), emergent, mature pines (1 m d.b.h.) and a lower, dense, secondary layer of young regenerating pines. In areas logged 15-20 years earlier the regenerating pole tree stratum often has 100 per cent crown closure, trees are 8-12 m tall and up to 15-20 cm d.b.h. The dominant tall tree crown closure is often less than 33 per cent.

In some areas a regenerating pine layer is absent: shrub chaparral communities (1-2 m high) or grass or sage vegetation form the understory (Plate I, Figure 3). This is more typical of gentler slopes and lower elevations in areas where there is less than 40 per cent crown closure. In other areas tall pines are absent and one or several layers or strata of lower pines are present. In yet others, regenerating pines form only a sparse cover. At low elevations or at higher elevations on exposed, west-facing slopes an admixture of pines and western juniper may occur (Plate I, Figure 4). Commonly dense shrubs of curlleaf mountain mahogany, 2-3.5 m high, form a narrow flank between such communities and openings.

Density and heights of vegetation layers, apart from indicating local physical and biotic habitat conditions, also reflect intensity of logging, frequency of cutting and time since logging activities ceased. In this region, the U. S. Forestry Service recognized four major size classes of forest based on tree diameters. These are (1) seedlings and saplings, up to 12.5 cm d.b.h.; (2) pole timber, 12.5-27.5 cm d.b.h.; (3) small sawtimber, 27.5-53 cm d.b.h.; and (4) large sawtimber, over 53 cm d.b.h. Heights of tree canopies for each of these classes, although variable, conform largely to the following: 1, up to 8m; 2, from 6-20m; 3, 15-30m; 4, 30-45m. Canopy density was classified into three classes of coverage of "crown-closure"—10-39%, 40-69%, and 70-100%. Within an area of 1.25 square kilometers of pine forest on the upper western slopes of Horsefly Mountain the U. S. Forest Service (1962) mapped over 20 different "timber type" communities, in an area of relatively uniform pine forest. Similar complexity is characteristic of the whole area. Young white fir often forms an important dense regeneration tree story, with pine or separately, at elevations above 1560 m.

Morain (1967) in his analysis of Horsefly Mountain forests in relation to radar imagery recognized the following four types as important:

1. Large patches of open (10-39% crown closure) "residual" forest 30-45 m high with natural regrowth of uneven age and height. Residual in this context refers to large, widely spaced old trees which have survived both the ravages of fire and logging. Some isolated patches of residual forest have no pine understory at all.

2. Mature secondary forest usually somewhat more dense than above, but not as tall. The pine understory usually contains even-aged trees either scattered or in patches, while beneath the pine there may be occasional shrubs of *Arctostaphylos* and *Ceanothus* from previous successional stages.

3. Immature pine forest with trees of pole timber size, 12-23 m. These vary in density up to 75 per cent or more crown closure and usually appear as a single-layered stand. They may also have a shrub understory.

4. Seedling and sapling stands. These occur generally as a layer beneath numbers 1 and 2 above. Where they do occur alone, they seldom have more than about 50 per cent crown closure and their height is commonly less than 10 m.

Lodgepole pine—On Yainax Butte, at elevations higher than 1950 m, lodgepole pine (*P. contorta*) replaces white fir and ponderosa pine on northern and eastern slopes. Lodgepole pine communities have not been logged and occur as dense, even-age, single-story stands of 10 m in height. Coverage is 100 per cent with very uniform canopy. The narrow band of lodgepole pine is difficult to distinguish from ponderosa pine on vertical panchromatic air photographs. It is more readily differentiated from the taller, darker, white fir communities. On the warmer southern and western slopes shrub communities of bitterbrush (*Purshia tridentata*) and curlleaf mountain mahogany (*Cercocarpus ledifolius*) occur above the ponderosa pine tree line.

White fir—At elevations over 1560 m white fir frequently occurs with ponderosa pine. In moist sheltered habitats, especially on northeastern slopes, white fir is codominant with pine. At elevations above 1650 m, on deep soils in areas sheltered from the south and the west, white fir becomes the dominant and frequently replaces pine. Where orographic precipitation results in high effective available moisture for plant growth, dense stands of white fir occur. This is the case on the northeastern slopes of Yainax Butte, Horsefly Mountain and Bly Ridge (Plate I, Figure 5). Because of favorable site characteristics on a steep northeastern slope west of Otto Boy Flat a dense stand of white fir occurs here at a lower elevation than elsewhere within the region (Plate III, Figure 4). This ridge site and adjacent Driscoll Springs flat will be discussed in some detail later.

Incense cedar—Structural and density variations are equally varied in the white fir communities as in the ponderosa pine. In addition, incense cedar (*Libocedrus decurrens*) occurs in small numbers in some mixed pine-fir communities (Plate I, Figure 6). Its distribution parallels that of the most favorable white fir habitats. Adult trees vary from 30-45 m in height. Trunk diameters can be over 1.5 m. It has been logged less than ponderosa pine, is less widely distributed, and does not occur in large single-species stands. This species has similar dense pyramidal form in its youthful growth to both white fir and ponderosa pine. If viewed from above, it does not differ greatly from either of these species in the adult stages. Viewed laterally the mature tree is more pyramidal than either ponderosa pine or white fir (Plate I, Figure 6).

Although juvenile and adult trees of white fir, ponderosa pine and incense cedar, when viewed from the ground, have distinct small differences in foliage, color, branching characteristics, and geometry or general form, these differences cannot be discerned on pan stereoscopic vertical air photos. The differences between pine and fir can be detected with difficulty and not always consistently on Ektachrome stereoscopic vertical air photos. The subtle differences between incense cedar, and either ponderosa pine or white fir cannot be clearly discerned on aerial color photographs. This is largely because the tree crown is pointed and small. Also incense cedar normally grows with branches closely interlaced with adjacent pine and/or fir.

Sugar pine—Small areas of sugar pine (*P. lambertiana*) occur intermixed with white fir, ponderosa pine and incense cedar on the eastern slopes of Yainax Butte. Although characterized by a very distinct stellate branching pattern they contribute little to the gross morphology of the communities in which they are found because of their scattered occurrence. Heights and trunk diameters of adult trees are similar to those of mature ponderosa pine and white fir.

Western juniper—Within the Horsefly Mountain region western juniper forms open woodlands chiefly on southern and southwestern slopes at lower levels than pine (Plate II, Figure 1). It also occurs in very dry eastern habitats between Bly and Tub Butte. It grows alone or frequently in combination with ponderosa pine. If pine and juniper are of similar height it is difficult to differentiate them from a distance. Adult trees may grow to 10-15 m in height. Unlike pines and fir mature trees retain the pyramidal form of youth. Commonly many open spaces occur between trees so total crown coverage is rarely above 39 per cent. These open areas are often rocky, supporting a low sparse cover of sage (*Artemisia* spp.) and rabbitbrush (*Chrysothamnus* spp.) Low sage and grass or low dry grass communities may occur where thin soils occur among the rocks.

Quaking aspen—Communities of quaking aspen, the only hardwood tree species in the survey area, have a very restricted distribution. Apart from an area of 8.5 hectares at 1800 m elevation on the northeast side of Horsefly Mountain (Plate III, Figure 3) quaking aspen is found only on narrow bands flanking stream edges or as very small stands in seepage hollows of flats. Because of the distinctive slender trunks,

whitish bark and decidious habit of these trees, they are locally conspicuous (Plate II, Figure 2). In autumn their bright golden autumn
foliage sharply differentiates them from other vegetation. Narrow tongues,
often of scattered individual trees, extend up damp valleys on northeastern slopes of Horsefly Mountain and less commonly on other northeastern slopes. Where they are intermixed with tall fir or pine they are
visible only from the ground.

The large area of hardwood, shown by Morain (1967) as growing on
the southeast slopes of Yainax Butte does not occur, and was wrongly
transcribed during his compilation of a vegetation map of the area. We
suspected this was not correct from the apparent anomaly of hardwood
in an area where habitat conditions appeared alien to its successful establishment. There is an extremely close relationship between habitat and
distribution of quaking aspen. Field checking verified that neither quaking aspen nor any other hardwood was present. Although quaking aspen
has probably the most widespread distribution of any tree in North
America it has a very narrow range of specific site requirements, especially near the ecological limits of its distribution.

Shrub Communities

Tall and low shrub forms and seral and climax shrub communities
are found within the Horsefly Mountain region. Taller shrubs are more
common in the chaparral shrublands; shorter shrubs predominate in the
sagebrush communities. Chaparral is usually a seral community in a fire
succession: sagebrush is often climax. In some cases sagebrush is a dis-
climax community representing a retrogression from tussock grass communities.

Chaparral—Early colonizers of forest burned areas are the sclerophyllous, medium height (1-1.5 m) greenleaf manzanita (*Actostaphylos
patula*) and tobacco bush or snowbush (*Ceanothus velutinus*). These
common species form an impenetrable scrub with a low, compact,
rounded canopy (Plate I, Figure 5 and Plate II, Figure 3). Coverage is
often 100 per cent on most areas that were known to have been burned
13 years ago. Coverage is less on more recently burned areas. The
established chaparral is a floristically rich community. Numerous dwarf
shrub species are found in spaces between the main bushes. Bitterbrush,
mountain mahogany, and thickets of bitter cherry (*Prunus emarginata*)
and western chokeberry (*P. virginiana* var. *demissa*) also occur. Bitter
cherry is found chiefly on damp slopes (Plate II, Figure 4).

Squaw carpet (*C. prostratus*), a prostrate spreading mat plant, is
common in open areas within forest along with the former two species.
These species form a nursery cover in which ponderosa pine and white fir
seedlings become established. As the trees grow and gradually achieve
increasing canopy cover these heliophilous or light-demanding chaparral
species die.

Rabbitbrush, especially rubber rabbitbrush. (*Chrysothamnus nau-
seous*), is common in the early stages of fire succession but becomes much

less important later (Plate II, Figure 3). This mesic, composite shrub is especially prevalent on disturbed areas such as road clearings, forestry logging sites, timber loading ramps, bulldozed tracks and earthwork dam margins. It also occurs with sage and is known as 'white sage.'

Rabbit sage, rabbitbrush, yellow sage or yellowbrush (*C. viscidiflorus*) is less common, occurring chiefly on dry slopes and benches. Other species of *Chrysothamnus* are present but less common than these two rabbitbrushes. Rabbitbrush shrubs are more characteristically associated with sagebrush than with chaparral.

Several taller shrubs, which appear after manzanita and snowbrush are established, frequently remain as a 2-3 m high shrub layer on the flanks of forest and around forest clearings when manzanita and snowbush have become much reduced. Curlleaf mountain mahogany (*Cercocarpus ledifolius*) appears the most euryphotic of these species. It is very common and is heavily browsed by deer to a psuedo-tree shape. All leaves, shoots and twigs within reach of the deer are eaten; the shrub can only attain a leafy canopy above the reach of the deer. Western mountain mahogany (*C. betuloides*), which is less common, is similarly shaped by deer browsing.

Bitterbrush, (*Purshia tridentata*), is another tall shrub that exists in the open but can also tolerate moderate shade. It is so heavily grazed by mule deer in the Horsefly Mountain region that it rarely reaches even 0.5 m in height. When it does escape beyond deer browsing height it forms a grotesque top-heavy bush (Plate II, Figure 4).

Curlleaf mountain mahogany, bitterbrush and mountain pink currant (*Ribes nevadense*), a spreading shrub that reaches to 1.5 meters, are common in small clearings in medium and tall forest with less than 50 per cent crown coverage (Plate II, Figure 5). These three species, especially the former two, are also common on exposed open southern slopes along with grasses. These euryphotic species represent a late state in the fire succession—burn, manzanita, bitterbrush-mountain mahogany, forest.

Sagebrush—Sagebrush (*Artemisia* spp.) shrublands are generally under 1 m in height. They vary in both height and density with habitat conditions. There is a particularly close relationship between soil depth, soil moisture and sagebrush communities (Plate I, Figures 1 and 4, Plate II, Figures 1 and 6, and Plate III, Figure 2). Normally the most stony plains are almost bare of vegetation. Stony areas with some thin soil support a dwarf shrubland of low or black sage, *Artemisia arbuscula*. Coverage is rarely 50 per cent and shrubs are rarely over 10 cm high. With a small increase in depth of soil small tussocks of grass such as Montana fescue, *Poa fecunda*, and squirrel tail (*Sitania hystrix*) appear along with dwarf shrubs (Plate II, Figure 6), and mat-forming, silt-binding plants such as *Anterinia*; much bare ground and rock are still present.

With increase in soil depth tall sage, *A. tridentata*, abruptly replaces black sage. Over most of the plains in the area there was scant evidence of an ecotone (transitional zone) between sage communities. Coverage or density of tall sage varied widely but the larger, taller (0.5-1 m) shrubs usually covered more total ground than did the dwarf sagebrush. Grasses are larger and more frequent. Numerous forbs including several showy composites may be present. Small rosette plants like yarro (*Archillia lanulosa*) and dwarf shrubs such as the spiny *Leptodactylon pungens* are common. Tall sage communities extend well within open pine forest areas (Plate III, Figure 1). They also cover open slopes if soils are not very shallow.

If the water table is near the surface silver sage communities replace tall or short sage (Plate III, Figure 2). Marked differences in sagebrush communities occur across Dry Prairie, Summit Prairie, Otto Boy Flat, Driscoll Spring, Gooch Meadow and other basins where there are but small changes in the micro relief. Changes in soil texture, depth and moisture do occur and these are reflected in vegetation community and density changes.

Grassland Communities

Three kinds of grassland are found in the Horsefly Mountain region: (1) remnants of bunch grass prairie with a closed continuous sod and grass tussocks 40-60 cm high; (2) moist meadows with a continuous, thick cover of grasses, sedges and herbs (often wild iris); and (3) sparse short grass often intermediate in position between moist meadows and sagebrush communities. It also occurs in many forest openings (Plate I, Figure 3).

The first of these communities was probably much more widespread prior to cattle grazing. It is restricted now to scattered areas of Dry Prairie forming small islands on ground several cm above the surrounding eroded sagebrush areas.

The mesic grasslands are localized about springs and also along seepage drainage channels where soil moisture is abundant (Plate I, Figure 1 and Plate II, Figure 2). They grade into true sedge marshes where standing water occurs. Narrow but sharp zonation of sedge, mesic meadow, drier grass and sagebrush can be observed flanking many of the small dams that have been constructed in this area.

The sparse, low cover and xeric nature of the third type of grassland reflects the drier conditions of its habitat. Ground cover varies greatly but it is often only 50 per cent. The presence of several species of prostrate forbs and marked soil cracking also characterize this community (Plate I, Figure 1).

IDECS PROCESSING OF MULTI-SENSOR IMAGES

The diverse plant communities discussed above occur with a wide range of sharply to diffusely-defined boundaries and with a wide range of densities within communities. The region thus lends itself well to the testing of image-combining devices now being used in vegetation studies by several groups in the U. S. Optical combining systems have been used by Colwell (1968) and associates in the University of California, Berkeley, and by Yost and Wenderoth (1967) at Long Island University. The system used at the University of Kansas is electronic rather than optical, and much of the investigation reported in this paper has centered around processing and evaluating images produced on this system which bears the acronym, IDECS.

The IDECS system, which is an Image Discrimination, Enhancement, Combination, and Sampling device now in its third generation, was designed and built at the Center for Research in Engineering Science of the University of Kansas. The IDECS system produces color images by combining data from up to six black and white multi-sensor images. Three groups of devices are used to achieve the color combinations: (1) Input devices (shown on the left in Figure 2) which are one source of information; (2) Control or processing devices (shown in Figure 3 and in the middle of Figure 2) which select the information desired; and (3) Output devices (shown on the right of Figure 2) which display the information.

In the following pages therefore we discuss the sequence of operations in the IDECS, starting with the image data and then proceeding through the Input, Processing, and Output devices.

Image Data

The color-combined images obtained with the IDECS were made with various combinations of 1:20,000 nine-lens multiband photography, and 1:11,000 color Ektachrome photography flown in October 1967 by the National Aeronautics and Space Administration (NASA). Panchromatic copies of the Ektachrome photos were made at a scale of 1:20,000 for use in the IDECS.

HH and HV radar polarization, 1:180,000, K-band radar images were flown in October, 1965, by Westinghouse Corporation for NASA. In both the HH and HV polarizations the radar energy is transmitted in the horizontal plane. Energy is reflected back in various planes, but only those which are horizontal contribute to the HH image and only those which are vertical contribute to the HV image. Because various objects appear differently in the two polarizations, additional information can be derived from two polarizations than from one alone. Thermal infrared (8-14μ) imagery with an azimuth scale of 1:75,000 also was obtained in October 1967. However, it was declassified too late for experiments in this study. Pan minus-blue vertical photographs (1:12,000), obtained in 1962, cover all but a tenth of the area. They were used in field and laboratory stereo studies.

Various combinations of the 1967 and 1965 images were used in the IDECS to produce 260 different sets of images on a color television screen. These were photographed on 35mm color film and carefully compared with the matching stereo photo pairs of both panchromatic and Ektachrome aerial photos and with a Forestry Timber-type map to determine if the boundaries shown correspond to differences in type and density of vegetation.

Color prints were made of 48 IDECS images selected for detailed

IDECS SYSTEM DIAGRAM

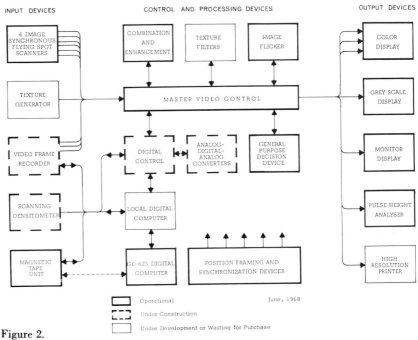

Figure 2.

field sampling. Both the color prints and transparencies were taken into the field. In addition many others were consulted for auxiliary information. Nine examples from this field-tested sample are evaluated later in this paper. In addition initial calibration tests of the IDECS are presented for two other sites.

Input Devices

Six image synchronous flying spot scanners—A flying spot scanner scans an image by moving a dot of light across the image along each of 525 parallel lines. By measuring the variation of intensity of the transmitted light at each instant, a signal is obtained in which time variation corresponds to position on the image. This signal can be transformed back to an image on display units, which operate like ordinary television receivers. The IDECS system is designed to simultaneously scan up to six images or photo-optically transformed images. By use of a synchronization pulse generator the dots of all six flying spot scanners at any instant are at exactly the same position on all scanning and display units.

Texture generator—The texture generator is an input device that generates standard patterns which will include crosshatch, vertical or horizontal slashes, a dot pattern—black dots on a white background or white dots on a black background—and controls to vary the size and spacing of dots. Different patterns will be used to indicate the location

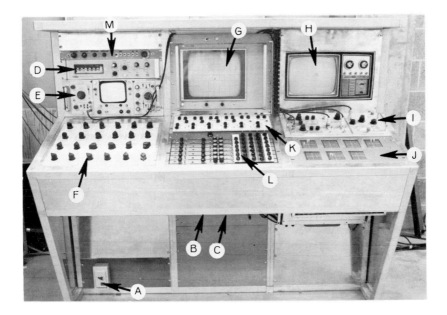

Figure 3. Part of the Image Discrimination, Enhancement, Combination, and Sampling (IDECS) System. Key — A) Master Power Switch, B) Synchronization Pulse Generator (this unit slaves displays and scanning units together so that all scan in unison), C) Power supplies (not visible), D) Counter, E) Time Base Generator and Raster Scope, F) Video Control Panel, G) Wideband Monochromatic Display, H) Color Monitor, I) Auxiliary Amplifier, J) General Purpose Decision Device, K) Positioning Control Panel, L) Video Processing Panel (includes Matrix, Level Select, Differentiation, and Gamma Correction), M) Analog to Digital Converter.

of many categories on map-like output images. This will supplement the present use of different colors to indicate categories.

Video frame recorder—The video frame recorder consists of a closed circuit video tape recorder which reads out data at high rates but records at slow rates. Used with the magnetic tape unit, which has a relatively slow output and a digital-to-analog converter, the video frame recorder allows computer calculations to be displayed.

Scanning densitometer—The scanning densitometer provides a high resolution input for use with the IDECS system.

Magnetic tape unit—The magnetic tape unit allows image data to be used in a computer.

Control and Processing Devices

Combination and enhancement—The combination and enhancement devices include a 6 x 4 matrix unit, level selection devices, and differentiation circuitry. All these may be fed in various combinations to the red, green and blue guns of the color display. Processing capabilities include:

straight color combinations of up to 6 images, the differences between 6 image pairs, edges enhanced in various colors, level selected portions of 6 images shown in color, and category identification and display.

Position framing and synchronization devices—The position framing and synchronization devices control the location, timing, and registry of images.

Texture filter—The texture filter employs a circular scan to perform localized harmonic analysis of points in a small area of the image. These harmonics are obviously related to texture. Not enough research has been done for us to know whether or not this will be a very valuable discriminant, but we suspect that it will at least be very good for discriminating linears from other systems. Texture discrimination if achieved with the texture filter will add appreciably to our ability to quantitatively detect differences between plant communities.

Master video control unit—This unit allows all useful interconnections between the inputs, processors, and the outputs to be linked in a great variety of ways to obtain optimal output.

Image flicker—The image flicker device alternately displays two images. This allows differences in tone and shape to be readily seen by either a flashing or an apparent motion. Uses of the image flicker include detection of (1) differences between images and (2) non-registry between images which the operator can correct with the position framing device.

General-purpose decision device (GPDD)—The general-purpose decision device takes the output from the six channel flying spot scanners and at each instant selects the most likely category for this output. This decision is presented in color thereby producing a map-like image. Suppose, however, that one wishes a more efficient output. A principal components transform of the input images could be implemented in the 6 x 4 matrix. Using this technique one is able to make best use of the resolution of the equipment. The principal components analysis from which the transform is obtained will come from the computer studies.

In the GPDD each input channel is divided into ten distinct levels. Since there are six channels, each with ten levels, this gives 10^6 possibilities in six-dimensional space. One can make decisions with the GPDD on the basis of most standard pattern recognition theories. It is possible to use it in two ways; that is, as a device to *make* decisions and as a device to *find* decisions. A program is now being written at CRES which will enable Bayes' decision theory (Wilkes 1962) to be programmed into the general purpose decision devices.

Digital control—The digital control is the digital circuit equivalent of the master video control. Digital control involves use of a local digital computer, and the University of Kansas GE-625 computer. The former manages the magnetic tape, the video tape recorder, the scanning densitometer and the analog-to-digital converter. Each of these has its own format and may need to be changed to make the data compatible with another portion of the system. The GE-625 will be used to perform statistical manipulation of data, and to provide inputs to the local digital computer and the general purpose decision device.

Analog-digital-analog convertors—These devices are used to convert analog data to digital format for computer processing, and conversely to transform digital data into an analog form for viewing or for image-formation.

Output Devices

Color display—The color display is a modified commercial color television set.

Gray scale display—This display is a black and white TV set that is used to display a single image, a portion of an image (level selection), or to show a single scanned line.

Monitor display—The monitor display shows the amplitude of the signal in time that is used to intensity modulate the gray level display or the color display. It is equivalent to a densitometer trace except that in the way it is presently connected it shows 525 densitometer traces piled on top of each other. One of the functions that we are working on will allow any one of the lines to be selected for densitometer work. The black and white TV will display the line selected for densitometry by intensifying that line on the black and white image. This device will also perform the following other functions:

(1) Calibration of all outputs. Calibration is achieved by scanning a grey-scale wedge of known density steps and setting the bottom and top of the wedge values to calibration marks on the face of the monitor. A nonlinear amplifier is now being built to compensate for logarithmic compression and to obtain linear steps with a density wedge.

(2) Reconstitution of previously obtained results employing level selection.

High resolution printer (HRP)—The high-resolution printer produces a high resolution gray scale processed image. Color outputs may be re-run as gray scale on the high resolution printer and then be combined optically with color filters to obtain high resolution color. Thus, with some sacrifice in time we can maintain resolution which is not feasible with the present color display.

The primary aim of this device is to improve versatility allowing more research problems to be simulated.

The HRP is a very slow but very high resolution intensity modulated cathode ray tube with mechanical attachments to expose $9\frac{1}{2}''$ sheets of film with about 30-40 line pair resolution per mm. Its principal use will not be with flying spot scanners because these are relatively coarse resolution devices. Rather the HRP will be used with the digital control and magnetic tape records of high resolution imagery. After various computed manipulations it will be feasible to print out with the HRP.

TEST SITE STUDIES

The following section discusses processed images obtained with the
IDECS system. These images cover seven test sites which were selected
from the many inspected to include the major vegetation communities and
densities in the Horsefly Mountain area.

Skedaddle Spring Test Site

This test site, with an average elevation of 1524 meters and an area
of approximately 18 square kilometers, is located five km southwest of
Bly, Oregon. A considerable part of this site on the northern slopes and
crest of Bly Ridge was burned over in 1955. The old burns are now
covered by a dense mixture of greenleaf manzanita (Plate II, Figures 3
and 4) and snowbush with scattered ponderosa pine. Forests in the test
site are predominantly ponderosa pine with some white fir and juniper.
Fairly large areas within the test site are bare or covered by sagebrush.

Explanation of Plate IV, Figure 1—One of the more useful IDECS
images of this area resulted from combining and enhancing two HV K-
band radar images. The IDECS system was manipulated to achieve the
effect of superposition of negative and positive images by adding colors
to one of the radar images and subtracting colors from the second.

The enhanced image thus produced recorded forests as yellow, shrub
communities as green, and bare or very sparsely vegetated areas as light
blue. Different densities of shrub vegetation were indicated by different
shades of green. Dense chaparral, as shown in Plate II, Figures 3 and 4,
showed as dark green and less dense chaparral as green. The different
growth form of dense sagebrush-covered clearings, similar to that seen in
the foreground of Plate III, Figure 1, was separated out as very dark
green. Mixtures of trees with chaparral were shown as orange. Thus both
structural and density variations within some of the vegetation were
demonstrated by this IDECS combination using only HV radar images.

Dry Prairie Test Site

The Dry Prairie Test Site includes approximately 42 square kilometers, consisting mostly of ponderosa pine forest on the hillsides and
sage and wet meadow plant communities of the Dry Prairie alluvial
basin (Plate III, Figure 2).

This large prairie, at an average elevation of 1190 m, contains low,
swampy areas with sedges and rushes, and slightly higher areas with
sagebrush and bitterbrush. Sagebrush communities on the hillsides, intermediate in soil type and elevation between the alluvial basin and the
hillside, have a distinct density that is detected both on the IDECS images
and on aerial color Ektachrome photographs.

Explanation of Plate IV, Figure 2—A further testing of image manipulation with two HV K-band radar images was carried out for the very
different vegetation patterns of Dry Prairie. Positive and negative images

were again superposed. In addition, the separate inputs from the channels were manipulated and enhanced to produce this image.

On this image wet, swampy meadows and the wetter stream channels (M), similar to those in Plate I, Figure 1 and Plate III, Figure 2, are shown in dark blue-green; higher, dry ground covered with sagebrush (S) appears in yellow; slightly higher ground covered with dense mixtures (B) of sagebrush and bitterbrush are shown as dark green surrounded by yellow; and ponderosa pine forests are colored orange-yellow (P) with more dense stands of trees shown in orange (D). Red denotes hillside areas composed of clearings covered with a mosaic of tall sagebrush (which has a different density from sage on the flat prairie) and scattered patches of ponderosa pine.

The area shown in red was not recognized as being different until the IDECS system indicated the difference. Aerial panchromatic and Ektachrome stereo-pairs confirm that a difference in sage vegetation density does exist.

The hillside clearings present a variety of aspects to the radar pulse, yet they are all red. We interpret this to mean that while the variations in return as a function of slope angle cannot be ignored, the variation in vegetation density documented here is largely responsible for the differentiation of the two sagebrush densities.

Horsefly Mountain Test Site

This 155 hectare test site lies on the east side of Horsefly Mountain at an average elevation of 1400 meters. A burned-over area is now windrowed and replanted with pine seedlings less than one-eighth meter in height. Mixtures of various densities of snowbush, greenleaf manzanita, and mountain mahogany cover much of the area, including portions of the windrows (Plate III, Figure 3). Mature ponderosa pines occur singly and in clusters in the test site. Considerable bare ground occurs throughout the site.

Explanation of Plate IV, Figure 3 — This figure was produced by combining three multiband images from the blue-green, orange-red, and near infrared portions of the spectrum. A density level showing bare ground was selected from the orange-red image and superposed on the combination of the three images.

Bare ground or areas with very little vegetation (including a road on the left), are shown in red. Blue-green denotes areas with low density shrub vegetation (S); low, dense chaparral is shown in olive green (D) as in the center of the image; and trees or taller shrubs are indicated by dark blue (T). The blend of colors should be noted as in parts of the windrowed area where the combination of bare ground and low densities of chaparral (B) produces a reddish color (cf Plate III, Figure 3 and Plate IV, Figure 3). The predominantly green color on the left hand side of the windrowed area (C) indicates a rather extensive cover of chaparral on the windrows.

Explanation of Plate IV, Figure 4 — This image was produced in the IDECS system by combining the HH polarization of radar, near infrared, and panchromatic photography. Superimposed on this image were density levels selected from the panchromatic to categorize vegetation types. By the process of differentiation on the panchromatic image the IDECS system outlined (in light green) the borders between areas of different vegetation density.

In the above figure, clusters of ponderosa pine are shown in red. Blue indicates ground that is bare or has only sparse vegetation; green indicates vegetation that is predominantly chaparral; the darker the green the more dense is the vegetation. The dark green area (D), in the center of the image, representing dense chaparral, and the light green patches (on the right) indicating less dense vegetation are clearly differentiated. Certain similarities with the results obtained for the Skedaddle Spring Test site using only HV radar are apparent.

Driscoll Spring Test Site

This test site is situated at an average elevation of 1,524 meters. A northeast-facing, 60 meter-high escarpment extends along the east side of the test site (Plate III, Figure 4). This escarpment supports a white fir forest, similar to that shown in Plate I, Figure 5, whereas the rather flat top of the ridge is covered by a two-story ponderosa pine forest (Plate I, Figures 2 and 4) with openings of various sizes. These forest openings contain mixtures of greenleaf manzanita, snowbush, mountain mahogany, mountain pink currant, sagebrush, and grass (Plate II, Figure 5). Although the larger clearings appear uniformly flat there are small differences in soil depth and moisture and in micro-relief. Minor depressions contain small marshes with sedges, rushes, wild iris, and grass. Areas only slightly higher, support short and tall sage (Plate III, Figure 1).

Explanation of Plate IV, Figure 5—This figure, covering 100 hectares, was produced in the IDECS system by combining the HH polarization of radar with two multi-band images (orange-red and near infrared). A contribution from the radar was the delineation of the forest openings. In this image various types and densities of vegetation are again represented by different colors. The light blue in the large clearings coincides mainly with wet areas with rushes, sedges, and grass but also includes some dirt roads. The darker blue indicates better-drained parts of the meadow (only about 5 cm higher). The green represents sagebrush with some grass. Shrubs and small trees are shown in green; tall ponderosa pines are indicated by reddish-brown, and the dense white fir is shown in a very dark red-brown color (F). It was thought that the dark coloring for the forest may result from radar shadow but checking of this and of vertical air photos showed that this darkness was indeed a density difference associated with a dense and relatively even canopy of the white fir stand. The larger forest openings are shown in blue, whereas smaller openings are shown in yellow. Because yellow borders appear at the edges of small and medium-sized clearings, the combined width of the yellow borders around small clearings is so great that blue does not show through the yellow. Therefore, the color of the clearing is an indication of its size.

Explanation of Plate IV, Figure 6—This image covers 80 hectares and combines near infrared, panchromatic, and the HH polarization radar. Forest openings are shown in various shades of blue (O). Tall ponderosa pine trees are shown in reddish-brown. The white fir forest on the northeast side of the ridge is shown in dark green (F).

A series of images of this test site, always using near infrared and HH polarization of radar, but different combinations of other multiband images all gave very good results. This suggests that the broad EM coverage given by near IR and K-band radar contributes valuable data for processing on the IDECS.

Yellow Jacket Spring Test Site

This test site of 155 hectares is one and one-half kilometers northeast of the Driscoll Spring Test Site. The same northeast facing, white fir-covered escarpment extends through both test sites (Plate III, Figure 5). This site differs from the Driscoll Spring site in having below the escarpment a large area with bare ground between chaparral and scattered ponderosa pines. The level area above the escarpment supports a ponderosa pine forest with many forest openings. This area has relatively little shrub cover between the trees or in the forest openings (Plate I, Figure 3).

Explanation of Plate IV, Figure 7—With the IDECS system panchromatic and near infrared were combined for a background image upon which a density level from the near infrared was superposed and shown in red. Brown indicates a greater influence of the level selection and yellow a lesser influence. Light green (as at letter G) indicates bare ground; yellow indicates ground with a sparse cover of shrubs; dark green (as in the lower left) indicates rather dense shrub cover between scattered ponderosa pine; brown indicates scattered ponderosa pine with very little brush between trees (P). The avocado green (at letter C) indicates an area of dense chaparral shrub. The white fir forest on the northeast-facing escarpment (F) is shown in very dark green with lighter green (at top) to indicate where the forest is less dense.

Explanation of Plate IV, Figure 8—For this figure ITEK images 3, 4, and 6 (from the green, yellow and red portions of the spectrum) were combined in the IDECS. The green lines, which represent lines of rapid change in density in the red-spectrum multiband image, outline the edges of the white fir forest and the dense patch of chaparral (at the letter C). Density levels (from the red-band image) that show (1) the white fir forest (F), and (2) the large areas of predominantly bare ground (B) are dominant features of this processed image.

Less dense areas of the white fir forest are shown in brown and more dense areas in dark green. Because this image categorizes vegetation by density the patch of dense chaparral shrub (C) is shown in a color of brown similar to that depicting the white fir. Examination of Plate IV, Figure 7, however, as a cross-check, separates this patch of chaparral from other types and densities of vegetation in the test site.

Little Blind Spring Test Site

This 155 hectare test site is located on the southeast side of Horsefly Mountain at an average elevation of 1,433 meters. The large clearings in the site (Plate III, Figure 6) have two types of vegetation: low sage grows on basalt rubble fields (Plate II, Figure 6) in the centers of the clearings whereas grass, similar to that seen at the left of Plate I, Figure 1, grows around the edges of the clearings where there are very few rocks on the surface. Dense chaparral shrub covers some hillsides, some areas are almost devoid of vegetation, and ponderosa pines cover other parts of the test site.

Explanation of Plate IV, Figure 9—This picture resulted from the combination of the orange-red and the near infrared multiband images. Densities were selected from both images and superposed on the image from the orange-red part of the spectrum. While this IDECS image doesn't show fine details it does categorize bare ground, sage-covered boulder fields, grassy areas, dense chaparral, and areas of ponderosa pine and shrubs.

In this image the grassy areas at the edges of the large clearings are shown in yellow; white indicates the low sage-covered basalt rubble fields (Plate II, Figure 6); blue indicates areas predominantly devoid of vegetation; green indicates ponderosa pines or a mixture of ponderosa pine and shrubs; and brown indicates dense chaparral.

Gooch Meadow Test Site

Located 4 to 5 km northwest of Horsefly Mountain Lookout at an elevation of 1300 meters this site occupies part of a low area extending from west of Yellow Jacket Spring to Horsefly Mountain. The broad vegetation pattern is one of well delineated isolated stands of ponderosa pine with intervening sagebrush, grassland and juniper communities.

In our analysis of this site we have used two adjacent areas of ITEK Multiband imagery for quantitative evaluation; one as a training set to make an IDECS matrix color combination; the other as a prediction set using the same IDECS settings. Each covers 155 hectares with ⅓ overlap.

The areas were selected on the basis of 1) comparability of vegetation categories, 2) slight relief, 3) sharpness of vegetation boundaries, and 4) simplicity of overall pattern. All vegetation categories in the training set are found in the prediction set but no additional categories exist. Four main types of vegetation are present; namely, 1) ponderosa pine forest with pine understory, 2) ponderosa pine/juniper mix, 3) sagebrush/bitterbrush shrubland, and 4) grassland, and two sub-categories: tall-dense/short-sparse sagebrush and dry/wet grassland.

Maps of each vegetation category and transparent overlays of boundaries were prepared for use on the IDECS. The transparent overlays for the training and prediction sets were used for congruencing the three images to be combined.

Two ITEK Multiband images were employed in each of the combinations: Frame number 6 ($\lambda = 635$ - ca. 700 mμ) and frame number 9 ($\lambda = 775$ - ca. 890 mμ) were electronically combined in both training the prediction sets with panchromatic photography converted from aerial Ektachrome and reduced to equivalent scale. Frames number 6 and 9 are reproduced in Plate V, Figure 1, and Plate VI, Figure 1.

Of the six major categories of vegetation mapped from field and photo analysis four somewhat broader yet internally consistent categories were displayed in the color combination of the training set (Plate V, Figures 1 and 2). In the combination of the prediction set this number reduced to three with the complete disappearance of the juniper-pine category (Plate VI, Figure 3).

Table 1 indicates the area percentages (measured by grid counting) detected by vegetation category in the training and prediction sets respectively. In the training set 70% of the pine forests were discriminated on the color combination against 55% on the prediction set. The prediction set contains a larger proportion of small isolated stands of ponderosa pine than the training set. Most of these are not resolvable with the relatively coarse resolution of the cathode ray tube, hence the lower identification percentage.

The remaining differences between vegetation categories in the training and prediction sets shown in Table 1 are more complex and no explanation of the differences will be given here. The results in the prediction set, however, do encourage further study of sample sites using the IDECS electronic system. Such studies are underway.

TABLE 1

VEGETATION CATEGORY	DETECTED	
	(Percent of area)	
	training set	prediction set
PINE FOREST	70	55
JUNIPER-PINE MIXTURE	53	0
SAGEBRUSH SHRUB	47	78
GRASSLAND	84	49

SUMMARY

Although ponderosa pine forests of varied age and density cover more than half of the Horsefly Mountain area in Oregon, there is a wide range of vegetation communities present. These have been mapped using several photographic and other forms of sensor imagery augmented by field survey.

Utilizing various combinations of multiband, radar, and Ektachrome photography electronic analyses with the University of Kansas' IDECS system were made of numerous areas. Techniques used included differentiation, level selection, combination, and texture discrimination for recognizing vegetation boundaries and density differences both within and between plant communities. Some 260 image combinations were displayed on a color TV screen and photographed on 35mm color film.

These photographs were compared with panchromatic and Ektachrome vertical aerial photos and a forestry map of the region. Color prints were made of a selection of 48 which were tested for accuracy of boundaries and densities in the field. The nine examples presented in this paper demonstrate in part the scope of the IDECS for multisensor analysis of vegetation. They are not necessarily the best combinations in terms of information content but were selected because they demonstrate some of the IDECS uses with further refinement of techniques. Experimentation with combinations of near infrared, thermal infrared, and radar imagery may yield instructive data.

Following our field investigations we considered some 26 of the 48 field tested IDECS images warrant further quantitative analysis. Testing of different areas with similar vegetation patterns to see if consistent results are achieved is presented for two sites.

Our studies demonstrate that in addition to IDECS discrimination of boundaries and densities in natural vegetation, combinations of imagery using the near infrared were usually better than those where it was not present. Also HV radar was useful for delimitation of openings in vegetation. Ready discrimination of major vegetation types—trees, shrubs, and bare or sparsely vegetated areas—and of combinations as well as of vegetation density differences from one K-band polarization suggest that radar can serve a useful function in vegetation mapping.

REFERENCES

Abrams, L. and Ferris, R. J., 1940. *Illustrated Flora of the Pacific States —Washington, Oregon, and California.* Stanford University Press.

Atwood, W. W., 1940. *The Physiographic Provinces of North America,* Ginn, Boston, 536 pp.

Colwell, R. N., 1968. Remote sensing of natural resources, Scientific American, 218:54-69.

Fenneman, N. M., 1931. *Physiography of Western United States,* McGraw Hill, New York, 534 pp.

Hunt, C. B., 1967. *Physiography of the United States,* Freeman, San Francisco, 480 pp.

Morain, S. A., 1967. Field studies on vegetation at Horsefly Mountain, Oregon, and its relation to radar imagery. CRES Report No. 61-22, Center for Research in Engineering Science, University of Kansas, Lawrence.

Morain, S. A. and Simonett, D. S., 1966. Vegetation analysis with radar
    imagery. Proc. of the Fourth Symp. on Remote Sensing of Environ-
    ment, University of Michigan, Ann Arbor: 605-622.

Morain, S. A. and Simonett, D. S. 1967. K-Band radar in vegetation
    mapping. Photogr. Engr., 33:730-740.

U. S. Forest Service, 1962. Timber-type Map, Fremont National Forest,
    Oregon, 1:31,680. Bonanza Quadrangle, no. 439.

Wilkes, S. S., 1962. *Mathematical Statistics*, John Wiley and Sons,
    New York.

Yost, E. T., and Wenderoth, Sondra, 1967. Multispectral color aerial
    photography. Photogr. Engr., 33:1020-1033.

EXPLANATION OF PLATE I

Figure 1.—Two-tiered ponderosa pine covers the background ridges; tall sage flanks the forest margins. Sharp boundaries, resulting from slight differences in soil depth, texture, and moisture content, separate tall sage from wet meadow (dark area on Summit Prairie), dry meadow (light even area), and low sage (in the foreground).

Figure 2.—A multi-tiered ponderosa pine forest with scattered, 45 m tall, emergent, adult pines. The dense secondary layer, about 5 m high, is mostly pine with some white fir. Scattered lower pines, 1-2 m tall, form a third discontinuous tier. This structure is typical of much cutover forest in the Horsefly Mountain region.

Figure 3.—Open tall ponderosa pine forest (foreground) with grass ground cover and lack of regeneration layer contrasts with multi-tiered, mixed ponderosa—white fir forest (in rear) with several dense regenerating layers. The former community is on flat ground; the latter is on flanking slopes.

Figure 4.—Mixed two-tiered ponderosa pine—western juniper community. The trees of the secondary layer are difficult to differentiate. The scattered tall emergent pines are no longer pyramidal in form as are the lower pines and the junipers and are readily distinguished. Some curlleaf mountain mahogany flanks the tall pine at the right as a band of tall shrubs. Gray colored, tall sagebrush forms the predominant ground-shrub layer.

Figure 5.—Dense even-age stand of white fir on a moist sheltered northeastern site above 1650 m elevation on Yainax Butte. One younger and lower (15 m) white fir tree is present (center) beyond the 2 m tall chaparral scrub of greenleaf manzanita and snowbrush.

Figure 6.—Three mature residual trees of incense cedar (left), ponderosa pine (center), and white fir (right) tower above a secondary, denser regenerating layer of young pines. Low chaparral scrub (foreground) covers an open area within the forest.

PLATE I

EXPLANATION OF PLATE II

Figure 1.—Western juniper woodland with low pyramidal trees. Tall sage forms the ground cover. Taller ponderosa pines with darker foliage are conspicuous at the left.

Figure 2.—The light-colored, broadleaf, deciduous foliage of a small stand of quaking aspen contrasts vividly with the dark, evergreen, needle-leaf foliage of ponderosa pine clothing the ridges beyond the alluvial basin. The quaking aspen and dark meadow grass (wet meadow) are associated with wet soils. Lighter, shorter, and less dense dry meadow replaces the latter on the drier margins of the alluvial basin.

Figure 3.—Dense, compact chaparral scrub, dominated by green-leaf manzanita and snowbush, has formed a seral nursery cover for establishment of ponderosa pines following burning 13 years ago. Rabbit brush (right foreground), an important initial colonizer of fired areas, is progressively replaced by sclerophyllous chaparral shrubs. Young pines have all grown since the fire. Some tall residual pines that escaped firing are visible on the skyline.

Figure 4.—Chaparral. The compact rounded canopy of dominant manzanita and snowbush shrubs on a 13-year old burn is broken by the taller, irregular shrub of bitterbrush (foreground) and by the larger clump of tall shrubs of western chokeberry and bitter cherry on a damp slope (center middle distance). The lower part of the bitterbrush shrub has been deformed by moderately heavy browsing by deer. The pines beyond the chaparral scrub mark the boundary of the fire.

Figure 5.—Small forest clearing with low shrubs of mountain pink currant, lower browsed bitterbrush, grass, and old partially burned logs.

Figure 6.—Low sagebrush. Low dwarf sagebrush and sparse grass growing in small soil pockets between boulders on a stony lava plain.

PLATE II

EXPLANATION OF PLATE III

Figure 1.—Tall sagebrush. Tall sagebrush forming the understory within an open ponderosa pine forest. The sagebrush is less continuous in the background multi-tiered forest where pine regeneration is well advanced.

Figure 2.—Sagebrush. Silver sage found on an area of damp soils (shown as darker in photo) on Dry Prairie replaces low sage on shallow dry soils in foreground and tall sage (lighter colored) beyond on deeper dry soils.

Figure 3.—Vertical aerial photo of Horsefly Mountain Test Site. Approximate scale:1:20,000.

Figure 4.—Vertical aerial photo (1:20,000) of Driscoll Spring Test Site. The light areas are Driscoll Spring Flat at the left and Otto Boy Flat on the right. The dark band of dense white fir forest along the northeastern flank of the ridge is also conspicuous.

Figure 5.—Vertical aerial photo (1:20,000) of Yellow Jacket Spring Test Site.

Figure 6.—Vertical aerial photo (1:20,000) of Little Blind Spring Test Site.

PLATE III

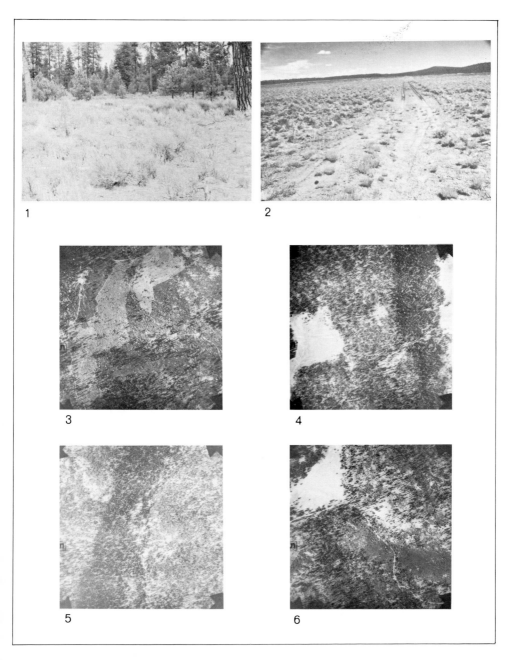

EXPLANATION OF PLATE IV

Figure 1.—Skedaddle Spring. Blue; bare areas or areas with sparse vegetation. Green; chaparral. Yellow; forests. Dark green (S); sagebrush clearings.

Figure 2.—Dry Prairie. Dark blue-green (M); swampy meadows and wet stream channels. Yellow (S); sagebrush. Dark green surrounded by yellow (B); mixture of sagebrush and bitterbrush. Orange-yellow (P); ponderosa pine forest. Orange (D); more dense stands of ponderosa pine. Red; hillside clearings with sagebrush density differing from that on the prairie.

Figure 3.—Horsefly Mountain. Red; areas bare or with sparse vegetation. Blue-green (S); low density chaparral shrub vegetation. Olive green (D); dense chaparral shrub. Very dark blue (T); trees and/or taller shrubs. Reddish color (as at B); combination of bare ground with scattered low density chaparral shrub. Greenish color (C); combination of rather dense chaparral shrub on bare windrows.

Figure 4.—Horsefly Mountain. Red; ponderosa pine. Blue; areas that are bare or have sparse vegetation. Green; chaparral shrub—the darker the green, the more dense the vegetation. Dark green (D); dense chaparral shrub.

Figure 5.—Driscoll Spring. Light blue; wet meadow. Brownish-green; shrubs and low trees. Dark blue; higher, drier meadow. Green in meadow; sagebrush with some grass. Reddish-brown; tall ponderosa pines. Very dark red-brown (F); white fir forest.

Figure 6.—Driscoll Spring (at a larger scale than Fig. 5). Blue; open areas. Yellow; shrubs and low trees. Reddish brown; tall ponderosa pines. Dark green (F); white fir forest.

Figure 7.—Yellow Jacket Spring. Light green (G); bare ground. Yellow; sparse cover of shrubs. Dark green (surrounded by brown as in the lower left); rather dense shrub cover between scattered ponderosa pines. Brown; areas of scattered ponderosa pine with very little brush. Avacado green (C); dense chaparral shrub. Dark green (F); white fir forest with forest less dense at top where green color is lighter and patterned.

Figure 8.—Yellow Jacket Spring. Dark brown (C); dense chaparral shrub. Very dark green (F) surrounded by brown at top center of image; white fir forest of moderate density—(F) at lower center; dense white fir forest. Pink (B); areas that are bare or with sparse vegetation.

Figure 9.—Little Blind Spring. Yellow; grassy areas at edges of clearings. White; basalt boulder fields covered by very sparse low sagebrush. Blue; bare areas. Green; ponderosa pine and/or shrubs. Brown; dense chaparral.

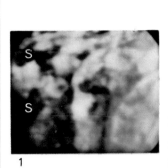

1

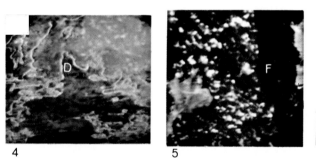

2

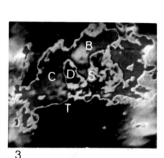

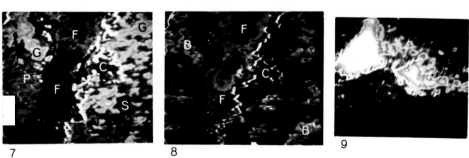

4

5

6

7

8

9

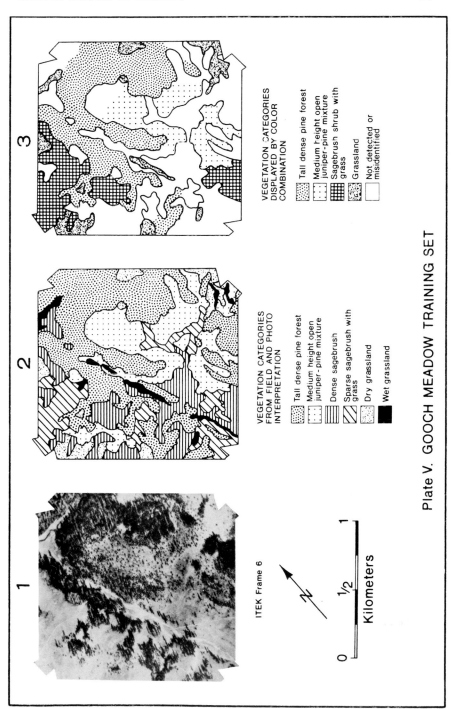

1

ITEK Frame 6

N

0   1/2   1
Kilometers

2

VEGETATION CATEGORIES
FROM FIELD AND PHOTO
INTERPRETATION

Tall dense pine forest

Medium height open
juniper-pine mixture

Dense sagebrush

Sparse sagebrush with
grass

Dry grassland

Wet grassland

3

VEGETATION CATEGORIES
DISPLAYED BY COLOR
COMBINATION

Tall dense pine forest

Medium height open
juniper-pine mixture

Sagebrush shrub with
grass

Grassland

Not detected or
misidentified

Plate V. GOOCH MEADOW TRAINING SET

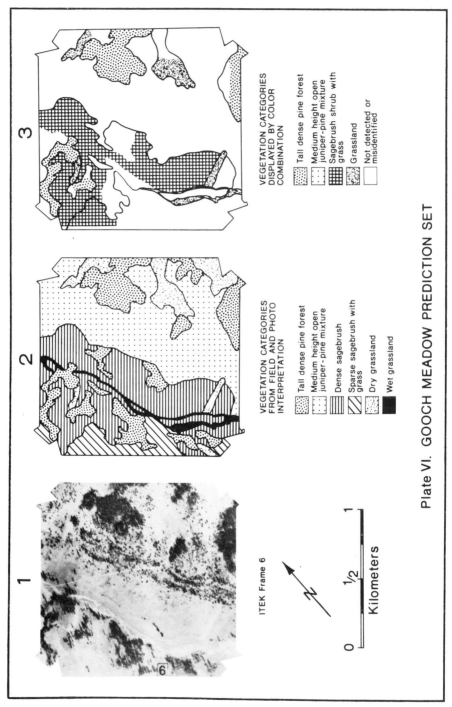

Plate VI. GOOCH MEADOW PREDICTION SET

6

Infrared Measurement of Plant and Animal Surface Temperature and Their Interpretation

DAVID M. GATES

INTRODUCTION

THE surface of a plant or animal is the transducer that couples the internal energy content and physiological response to the environmental factors. Energy flows between the surface and the environment and between the plant or animal interior and the surface. The degree to which the energy content of the plant or animal is coupled to any environmental factor depends upon the properties of its surface. An organism may not receive more energy than it gives out over an extended period of time or it will become warmer and warmer and perish. Nor can it give out more energy than it receives or it will get colder and perish. At equilibrium the surface temperature of the organism adjusts until the energy flowing in equals energy flowing out. Whenever any environmental factor changes in value or the organism changes its properties or relationship to the environment, the surface temperature adjusts to a new equilibrium value. In this chapter, steady state situations only are considered. Plants and animals do spend a fair percentage of their life history in momentary transient energy states. Eventually a careful analysis of transient states must be made.

The streams of energy flowing between a plant or animal and its environment are shown in Figure 1. There are many streams of radiation from many directions. There is moisture exchange in the way of sweating or transpiration. There is wind which causes convective cooling or heating. Even in its simplest context, the environment is complex.

ENERGY EXCHANGE FOR ANIMALS

In order to achieve an approximate analysis of the energy budget of an animal, one considers an animal as a series of concentric cylinders, as shown in Figure 2. The inner cylinder is the body cavity at a tempera-

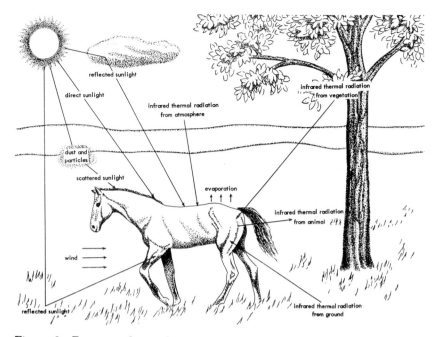

Figure 1. Energy exchange between an animal and the environment showing
   radiation streams, convection by wind, and evaporative cooling.

ture T_b where the metabolic energy M is generated and from which energy
E_{ex} is ejected by respiratory moisture loss. The net energy $(M - E_{ex})$
generated within the body cavity is conducted across fatty tissue of
thickness d_b to the skin surface at temperature T_s, which represents the
surface of the second concentric cylinder. The animal may sweat at the
skin surface and lose from it an amount of energy E_{sw}. Hence the net
energy from within the animal to be transferred from the skin surface is
$(M - E_{ex} - E_{sw})$. If the animal has fur or feathers, then this is represented
by a third concentric cylinder of thickness d_f and surface temperature
T_r. It is the surface temperature T_r which responds to the energy flow
between the animal and the environment. The conductivity of the fat is
K_b and the conductivity of fur or feathers is K_f. The ability of an animal
to maintain a certain temperature differential between its body temperature and its radiant surface temperature T_r depends upon the insulative
quality of its fat and fur or feathers. It also depends upon its metabolic
rate and moisture loss rate. The temperature differential between the
body cavity and the external radiation surface is the sum of the temperature difference across the fat plus that across the fur or feathers. Hence,

$$T_b - T_r = (T_b - T_s) + (T_s - T_r)$$

$$= \frac{d_b (M - E_{ex})}{K_b} + \frac{d_f (M - E_{ex} - E_{sw})}{K_f}$$

(1)

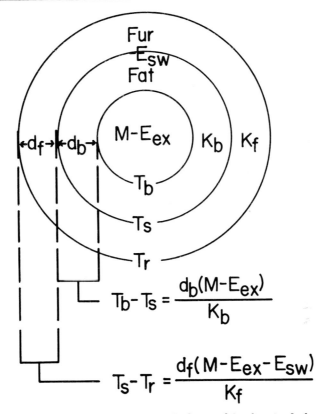

Figure 2. Schematic diagram of concentric cylinder model of animal for heat transfer considerations. The body cavity at temperature T_b is where the metabolic energy M is generated, surrounded by fat of conductivity K_b, thickness d_b terminating at the skin at temperature T_s, and which may be surrounded by fur or feathers of conductivity K_f, thickness d_f terminating at an external surface at temperature T_r. Respiratory loss of moisture is E_{ex} from the body cavity and sweating at a rate E_{sw} occurs from the skin.

Here one can see explicitly the role played by each of the animal properties in maintaining a certain body temperature relative to the radiant surface temperature of the animal. If the animal has neither fur nor feathers, then the radiant surface temperature becomes the skin temperature and the second term of Equation (1) drops out. Illustrations of the temperature differentials which various animals can maintain between body temperature T_b and the "radiant" surface temperature T_r are

shown in Figures 3 and 4. These are based on combinations of maximum
and minimum values of M, E_{ex}, E_{sw}, d_b, d_f, K_b, and K_f to give maximum
and minimum limits to the evaluation of Equation (1).

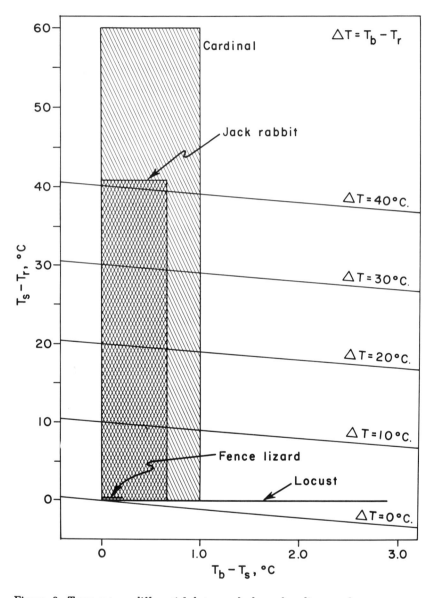

Figure 3. Temperature differential between body and radiant surface temperature
for animals. Abscissa is temperature difference across the body fat. Ordinate
is temperature difference across the fur or feather layer.

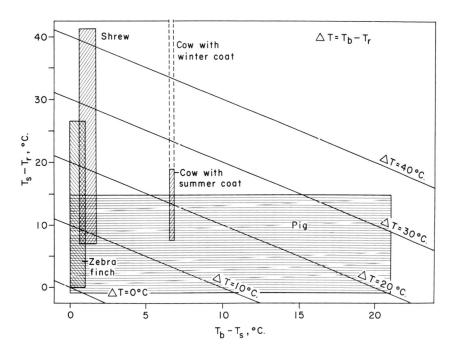

Figure 4. Temperature differential between body and radiant surface temperature for animals. Abscissa is temperature difference across the body fat. Ordinate is temperature difference across the fur or feather layer.

The energy budget of the surface of an animal is given by the following, where all terms are in $cal\ cm^{-2}\ min^{-1}$:

$$Q_{abs} + M = \epsilon \sigma T_r^4 + h_c(T_r - T_a) + E_{ex} + E_{sw} + C \qquad (2)$$

where Q_{abs} is the average radiation absorbed by all surfaces of the animal, ϵ is the emissivity of the animal surface to longwave radiation, δ is the Stefan-Boltzmann constant, h_c is the convection coefficient and C is energy lost or gained by conductance to the substrate upon which the animal is resting. The amount of radiation absorbed is complicated. The animal receives an amount of direct sunlight, S, scattered skylight, s, reflected sunlight and skylight, $r\ (S + s)$, and thermal radiation from the atmosphere, R_a, and from the ground surface, R_g. The animal presents certain surface areas, A_i to these various streams of radiation and each surface has a mean absorptivity, a_i, to each radiation stream. Hence, Q_{abs} averaged over the total surface area, A, of the animal is given by:

$$AQ_{abs} = \bar{a}_1 A_1 S + \bar{a}_2 A_2 s + \bar{a}_3 A_3 r (S + s) + \bar{a}_4 A_4 R_a + \bar{a}_5 A_5 R_g \qquad (3)$$

Each of the radiant streams, R_i, has a spectral quality which can be described monochromatically λR_i. The animal surface has a spectral absorptivity λa_i, which depends upon the wavelength λ. The average absorptivity a_i to each stream of incident radiation is given by:

$$\bar{a}_i = \frac{\int_{\lambda_1}^{\lambda_2} {}_\lambda a_i \; R_i \; d\lambda}{\int_{\lambda_1}^{\lambda_2} {}_\lambda R_i \; d\lambda} \tag{4}$$

The forced convection coefficient for a cylinder in crossflow is given by:

$$h_c = 6.2 \times 10^{-3} \; \frac{V^{1/3}}{D^{2/3}} \tag{5}$$

where V is the wind speed in cm sec^{-1} and D is the diameter of the cylinder. The dimensions of the coefficient are chosen so that h_c has the units of cal cm^{-2} min^{-1} °C^{-1}. If the cylinder is in longitudinal flow, a different coefficient applies; see Gates (1962). A cylinder in free convection has the following convection coefficient:

$$h_c = 6.0 \times 10^{-3} \left(\frac{T_r - T_a}{D} \right)^{1/4} \tag{6}$$

It is seen from Equation (2) that the surface temperature of an animal is a complicated function of many environmental factors and animal factors. The surface temperature as the dependent variable is a function of about six or more independent variables. This is a complex space in its simplest form and makes the problem of understanding the surface temperatures of animals difficult. Nevertheless, definitive answers can be achieved, as is now shown. From Equation (2) one can immediately understand the influence of radiation, wind, or air temperature on the surface temperature of any animal whose properties are known.

CALCULATED SURFACE TEMPERATURES FOR ANIMALS

A few explicit examples of the calculation of energy exchange for the surfaces of animals are given based on Equation (2). For this purpose a poikilotherm, the lizard, and a homeotherm, the cardinal, were selected. The body temperature for the lizard is always very close to its radiant surface temperature, since the insulation by the thin layer of fat is very small. The lizard will die if its body temperature exceeds 42.5°C. The lizard becomes relatively inactive when its body temperature exceeds 38 or 39°C. For the cardinal, the body temperature must remain very close to 42°C under all conditions in order for the bird to survive. The basic properties for each animal are given below.

	Lizard		Cardinal

Lizard
$M = 0.008 \text{ cal cm}^{-2} \text{ min}^{-1}$
$E_{ex} + E_{sw} = 0.003$
$D = 1.5 \text{ cm}$
$a = 0.8 \text{ to } 0.9$

Cardinal
$M = 0.077 \text{ (BMR)}$
$= 0.175 \text{ (Max)}$
$E_{ex} + E_{sw} \text{ (Min)} = 0.007$
$E_{ex} + E_{sw} \text{ (Max)} = 0.05 \text{ to } 0.08$

The following two environmental situations are used:

Sunny Day
$T_a = 30°C$
$Q_{abs} = 1.4 \text{ cal cm}^{-2} \text{ min}^{-1}$
$V = 10 \text{ cm sec}^{-1} \text{ (still air)}$
$= 100 \text{ cm sec}^{-1} \text{ (breeze)}$

Cloudy Day
$T_a = 30°C$
$Q_{abs} = 0.7 \text{ cal cm}^{-2} \text{ min}^{-1}$
$V = 10 \text{ cm sec}^{-1} \text{ (still air)}$
$= 100 \text{ cm sec}^{-1} \text{ (breeze)}$

The surface temperatures of the lizard and cardinal for each of the conditions specified above are given in Table II. It is obvious that the lizard cannot remain in the direct sun at an air temperature of 30°C with low wind speeds and would in fact seek shaded habitats. On an overcast day, the lizard could be quite comfortable in the open or elsewhere. The cardinal is able to maintain a body temperature 2 or 3°C below his surface temperature under hot conditions by reducing his metabolic rate and dumping as much water as possible. Gates (1968) has described the range of temperature differential possible for various animals. The cardinal is able to maintain an enormous temperature differential between body and surface during cold winter conditions by turning up his metabolic rate, improving his insulation by fluffing his feathers, and by releasing as little moisture as possible. It is evident that the cardinal cannot withstand the bright sun with no or little wind because of the high body temperatures which would result. One can elaborate on these calculations and show precisely the conditions which the cardinal can withstand.

Table 1. The calculated surface temperatures of a lizard and a cardinal for various environmental conditions representing a sunny day and a cloudy day, both at an air temperatureof 30°C.

		Lizard				Cardinal	
Q_{abs} cal cm^{-2} min^{-1})		1.4	1.2	1.0	0.7	1.4	0.7
V (cm sec^{-1})	10	65	56	46	30.5	75(1)	30(1)
						85(2)	30(1)
							36(3)
V (cm sec^{-1})	100	53	46	40	30	72(2)	
						67(1)	

(1) $M = E_{ex} + E_{sw}$
(2) $M = M_{max}, E_{ex} = E_{sw} = 0$
(3) $M = M_{min}, E_{ex} = E_{sw} = 0$

RADIATION THERMOMETRY

All surfaces emit radiation proportional to the fourth power of the absolute temperature of the surface. The intensity of the radiation depends upon the emissivity of the surface. In principle, one can measure the surface temperature by measuring the quantity of radiation emitted if the emissivity of the surface is known. Fortunately, the emissivity of most biological materials is between 0.95 and 1.0. Gates (1968) and Fuchs and Tanner (1966) have discussed the errors in the temperature interpretation caused by emissivities less than 1.0. The spectral composition of the radiation emitted by surfaces at ambient temperatures in the vicinity of 300°K is entirely of infrared wavelengths, with the peak intensity at about a wavelength of ten microns. An infrared radiometer or thermometer is used to detect the radiation emitted by surfaces whose temperatures are near ambient temperature. The radiation thermometer is calibrated to read the "apparent" surface temperature directly by means of a blackbody radiation standard.

Infrared radiometers of various designs have been available for some time and were described by Gates (1962). Stoll and Hardy (1952) designed a portable radiation thermometer, which was excellent for clinical use but had severe limitations when used in the field during rapid transient conditions. Gates (1961 and 1963) reported observations of leaf temperatures and the measurement of environmental radiation fluxes using the Stoll-Hardy radiometer. The Stoll-Hardy radiometer is a d.c. instrument subject to drift caused by temperature changes of the instrument.

Several prototype models of a portable infrared radiometer designed by the staff of Barnes Engineering Company have been available to the author for use during the last year. The instrument is described in detail by Gates (1968). We have used the portable radiation thermometer shown in Fig. 5 to measure plant and animal surface temperatures within many ecosystems, including alpine tundra, desert, rain forest, grasslands, and others in many parts of the world. The radiation thermometer has a thermopile detector which receives the radiation focused on the "hot" junctions while the "cold" junctions are at the ambient temperature of the instrument. There is an electronic chopper to interrupt the electrical output, which is then amplified by an a.c. amplifier and read out on a meter mounted on the rear of the instrument.

ANIMAL SURFACE TEMPERATURES

Table 2 is a listing of the surface temperatures of various cattle in the sun at the St. Louis stockyards on 6 June 1968. It is evident that there are substantial differences in surface temperatures of animals related to their hair color and absorptivity to sunlight. The visual appearance of the hair color is not always a good indication of the average absorptivity throughout the whole spectrum.

In Colorado last summer (1967) we measured the surface temperatures of a black dog and a white dog, each with thick fur. The incident flux of sunlight was very intense, or approximately 1.6 cal cm^{-2} min^{-1}. The black dog's surface temperature was at 45 to 53°C and the white

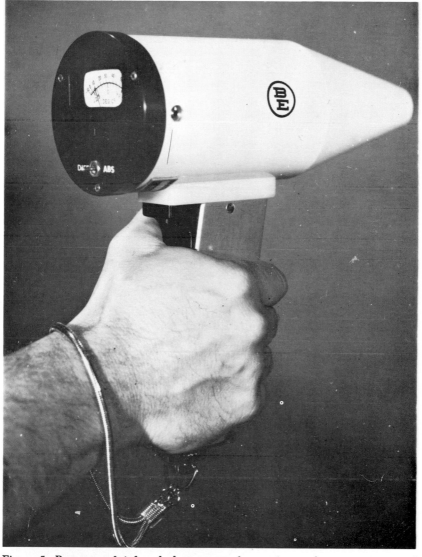

Figure 5. Prototype of infrared thermometer for sensing surface temperatures of plants and animals and for measuring the infrared radiation fluxes of the environment.

dog's at 30°C to 42°C when the air temperature was 22°C. During cloudy conditions the black dog's surface temperature was at 35°C and the white dog's at 30°C with air temperature 22°C. A black laborador in St. Louis in the sun had a maximum surface temperature of 32°C; most sunlit parts were at 25°C and shaded parts were at 5°C when air temperature was -20°C. The same dog in a wind of 3 to 5 mph at an air temperature of -12°C had all surfaces at -12°C. Equation (2) shows that the sur-

Table 2. Surface temperatures of cattle in sun at St. Louis Stockyards 6 June 1968. Air temperatures = 33-34°C, wind speed variable 0.6 to 2.5 mph, relative humidity 47%, downward shortwave radiation = 1.4 cal cm^{-2} min^{-1} at 1440.

	T_r °C	T_{soil} °C	Time
Hereford Back			
Brown	44-45	57	1430
White	39-40	57	1430
Red	44	57	1430
White	32	57	1441 wet fur
Red	40	57	1441
Black	43	57	1441
Brown	45	54	1444
White	34-35	54	1444
Black	46	54	1445
Hereford Belly			
Red	39-40	56	1435 Several animals
Black	40	56	1435 dry soil
Black	36	23	1435 wet soil

face temperature of an animal is a function of the metabolic energy generated and the moisture loss as well as the amount of radiation absorbed, the wind speed, and the air temperature.

Surface temperatures of many other animals have been measured using the portable radiation thermometer during many types of environmental conditions. The surface temperatures of lizards and snakes are generally very close to the subtrate surface temperature; see Gates (1968). The surface temperatures of birds in wind are usually very close to air temperature. The surface temperatures of large insects, particularly orthoptera, were measured and found to be near substrate temperature. A bumblebee on a penstemon flower was at 22°C, with flower at 19°C and air tempertaure at 20°C. A cicada was 0.3°C warmer than the grass at 22°C in very weak sunlight. The back of a wallaby in the sun in Australia was 43°C and in the shade was 32°C, with air temperature at 28°C. The back of a wombat in the sun was 44 to 47°C when the air temperature was 28°C and a Koala bear in the shade was at 35°C.

One of the most interesting applications of the infrared thermometer was the measurement of the surface temperatures of a red deer when the air temperature was 15°C during still air overcast conditions. Body surface temperature was 17°C, head surface temperature 25°C and antler surface temperature at 28-30°C. The surface temperature of antlers in velvet of a female reindeer at the St. Louis Zoo in a shed at 17°C was 33°C. The antlers of some ruminants are effective radiators and dump a significant amount of energy during the warmer part of the rutting season.

Surface Temperatures of Plants

The energy budget of a plant leaf takes much the same form

as the energy budget of an animal surface and is given by the following equation:

$$Q_{abs} = \epsilon \sigma T_\ell^4 + k_1 \left(\frac{V}{D} \right)^{1/2} (T_\ell - T_a)$$

$$+ L \frac{{}_sd_\ell (T_\ell) - r.h. \cdot {}_sd_a(T_a)}{r_\ell + k_2 \dfrac{W^{0.2}D^{0.35}}{V^{0.55}}}$$

where D is the dimension in cm of the leaf in the direction of wind flow, W is the dimension transverse to the wind, r is the internal diffusion resistance of the leaf in cm sec^{-1}, r.h. is the relative humidity, ${}_sd$ $(T$ $)$ and ${}_sd_a$ (T_a) are the saturation moisture densities of the air within the leaf at the leaf temperature T and in the air beyond the leaf boundary layer at temperature T_a. L is the latent heat of vaporization of moisture (580 cal gm^{-1} at 30°C). The following values are specified for the coefficients, as determined from many measurements within a wind tunnel:

W >> D	W << D
or	or
W = D>5 cm	W = D≤5 cm
$k_1 = 10^{-2}$	$k_1 = 1.6 \times 10^{-2}$
$k_2 = 3.5 \times 10^{-2}$	$k_2 = 2.6 \times 10^{-2}$

The leaf temperature is a function of the radiation absorbed by the leaf, the air temperature, the wind speed, and the water vapor pressure of the air. The transpiration rate for the leaf not only depends on the internal resistance of the leaf to the flow of water vapor through the stomates, but also depends upon the energy available for the vaporization of water. Leaf temperatures are more sensitive to relative humidity than are most animal surface temperatures, with the possible exception of amphibians.

An illustration of calculated leaf temperatures and transpiration rates for various air temperatures is given in Fig. 6 for a leaf of dimensions 5 x 5 cm, an amount of radiation absorbed of 1.0 cal cm^{-2} min^{-1}, a wind speed of 200 cm sec^{-1}, and a relative humidity of 50%. At moderate and low air temperature, the leaf temperature is always greater than air temperature, but at high air temperature the leaf temperature may be lower than air temperature when there is sufficient transpiration.

When the leaf dimensions is very small, such as with needles, the leaf temperature is very close to air temperature. This fact was established to be a great advantage to desert plants with low water availability; see Gates, Aldefer, and Taylor (1968).

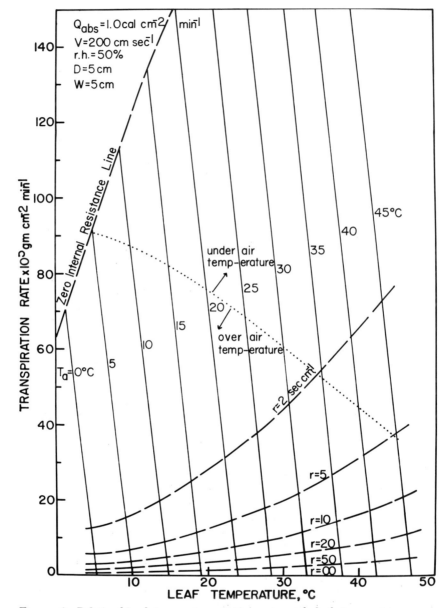

Figure 6. Relationship between transpiration rate and leaf temperature as a function of the air temperature and the internal diffusion resistance of the leaf at an absorbed radiation of 1.0 cal cm⁻² min⁻¹, a wind speed of 200 cm sec⁻¹, a relative humidity of 50%, for a leaf of 5 x 5 cm dimension. Many other diagrams of this type could be shown with each environmental factor as a variable.

Various plants side by side will have strikingly different temperatures. An example is given in Table 3 for a sunny but hazy day in St.

Louis, Missouri. The air temperature was 31°C, the relative humidity was 65% and the amount of incident sunlight and skylight was between 1.1 and 1.3 cal cm^{-2} min^{-1}. Leaf temperatures of various sunlit plants varied from a high of 48°C for the ginger to a low of 33°C for corylus. Shade leaves were from 29°C to 37°C, with several about 32°C. The young leaves of *Mahonia repens* were at 33 to 36°C, the mature older leaves at 40°C, and a senescent leaf at 42°C.

Table 3. Leaf temperatures of various sunlit plants at the Missouri Botanical Garden, St. Louis, Missouri, at 1200, 15 June 1968. Air temperature, 31°C relative humidity, 65%; and incident sunlight plus skylight between 1.1 and 1.3 cal cm^{-2} min^{-1}. Still air. Wet soil was at 40°C and dry leaves on ground at 57° C.

Canna	42-45, 48	Crab Grass	38
Magnolia x Loebneri	42	*Convulvus*	37
Poa (Bluegrass)	42	*Bumelia*	36
Aristolochia	42	*Berberis*	36
Young leaves	39	*Sonchus*	34
Mahonia repens	40	*Cedrus leboni*	34
Young leaves	33-36	*Corylus*	33
Scenescent	42		
Juniperus	39		
Plantago	38		

Acknowledgement

The author wishes to thank Dr. Warren Porter for many of the measurements of animal surface temperatures and for calculations of the energy budgets of animals. Mrs. Laverne Papian was responsible for much of the computer programming. This research was supported by the Center for the Biology of Natural Systems, Washington University, St. Louis, Missouri, under PHS Grant No. 1 P10 ES 00139-03 ERT and the computer facilities of Washington University under NSF Grant No. G 22296.

REFERENCES

Fuchs, M. and C. B. Tanner. 1966. Infrared thermometry of vegetation. Agron. J. 58:597-601.

Gates, D. M. 1961. Winter thermal radiation studies in Yellowstone Park. Science 134:32-35.

Gates, D. M. 1962. *Energy Exchange in the Biosphere*. Harper and Row, Publ. N. Y. 151 pp.

Gates, D. M. 1963. Leaf temperature and energy exchange. Archiv. Meteor., Geophysik u Biokl., Series B. 12:321-336.

Gates, D. M. 1968. Sensing biological environments with a portable radiation thermometer. Appl. Opt. 7 (In Press).

Gates, D. M., R. Alderfer, and E. Taylor. 1968. Leaf temperatures of desert plants. Science 159:994-995.

Stoll, A. M. and J. D. Hardy. 1952. A method for measuring radiant temperatures of the environment. J. Appl. Physiol. 5:117-124.

Analysis of Environmental and Vegetative Gradients in Yellowstone National Park from Remote Multispectral Sensing

Lee D. Miller and Charles F. Cooper

INTRODUCTION

Thermal infrared imagery has been used to detect a variety of hydrothermal areas in Yellowstone National Park (McLerran 1965). The dynamic nature of these general hydrothermal areas has also been recognized (Marler 1964). Ecological investigations of these changing areas prior to this study had concentrated primarily upon the algal communities of thermal springs and hot pools (Brock 1967a & 1967b). Hydrothermal activity is not necessarily associated with hot springs, mud pots or other direct surface manifestations. Within areas of extensive hydrothermal activity it is not unreasonable to expect zones of higher than normal soil temperature resulting from anomalously high terrestrial heat flux. Soil temperature is an important environmental factor in determining vegetative cover and any significant deviation from the normal heat flux must result in a new ecological balance.

In 1966 the senior author reported upon the successful use of thermal infrared scanner imagery and aerial photography to locate areas of anomalously hot earth in Yellowstone Park (Miller 1966a). Prior to that date one or two areas consisting exclusively of anomalous terrestrial heat flux had been observed by those familiar with the hydrothermal areas of Yellowstone National Park. Elsewhere, a large soil thermal anomaly has been documented near Lordsburg, New Mexico (Kintzinger 1956). Numerous hot earth areas have been studied in detail in the New Zealand hydrothermal areas (Benseman 1957, 1959a, & 1959b). In New Zealand the term *steaming ground* is used in referring to such areas as the anomalous heat results from steam diffusing upward through the soil (Elder 1965). Such steam diffusion also occurs in the anomalously hot earth areas documented in Yellowstone and the name *steaming ground* has therefore been adopted.

A ground investigation of the ecological relationships existing in the vegetation occurring on steaming ground was begun in 1966. Steaming ground sites in the Park are especially interesting from an ecological viewpoint in that they are dynamic in nature; they appear and disappear; they gradually change their energy budget; and they shift in their ground positions. This dynamic nature of the geophysics of the sites is being studied in terms of its effect on the vegetative cover of steaming ground and its potential documentation and study with remote sensing techniques.

New, up-to-date, multispectral imagery was acquired during the fall of 1967[1] for steaming ground areas previously located with earlier imagery. While the interpretation and processing of this new calibrated imagery has just been undertaken it has proceeded far enough to establish that quantitative information can be extracted. The thermal infrared imagery recorded on magnetic tape is being quantitized electronically into known temperature levels and an image produced. These same tapes are being further processed to produce contour maps of the apparent surface temperature of steaming ground and surrounding areas. Preliminary processed images correlated well with apparent temperature maps and related measurements made on the site at the time of the flight.

LOCATION AND DESCRIPTION OF STEAMING GROUND

The original imagery, obtained in May of 1961 and used for locating the study areas[2], was acquired simultaneously with an aerial camera using panchromatic film and an optical-mechanical scanner operating in the thermal infrared (McLerran 1965). At that time there were still large patches of snow remaining on the ground, while some hot springs and pools in the Park were at the boiling point, which at that altitude is approximately 92.5°C. The dynamic range of the thermal imagery encompasses this entire range of ground temperatures. The imagery is thus essentially a hot versus cold surface temperature map. The resultant lack of temperature discrimination permitted development of only a relatively simple interpretation scheme.

The scheme which evolved reflected the criteria which imaged hot earth sites had to meet to subsequently allow meaningful ecological interpretation. The anomalous terrestrial heat flux had to be manifested at the ground surface as a higher temperature only and not as hot water, escaping hot gases, or hot mud. The presence of any of these factors would have added additional complicating environmental variables to the ecology

---

[1]This image block was acquired and is being processed for the U. S. Geological Survey by the Infrared and Optical Sensor Laboratory of the University of Michigan.

[2]This image block was acquired for the Cold Regions Research and Engineering Laboratory of the U. S. Army by the Infrared Physics Laboratory of the University of Michigan.

of the site. It was also important that the selected areas were well-drained and adjacent to major vegetation types, especially lodgepole pine *(Pinus contorta)*, the dominant tree species in the lower elevations of the Park.

The geothermal areas within the Upper, Midway and Lower, Geyser Basins of the Park were clearly portrayed on the infrared imagery. The vast majority of these were rejected in the interpretation phase as probable direct, surface, hydrothermal manifestations. The decision as to whether a hot area was a hot spring or pool, mud pot, gas vent, or the desired steaming ground was based upon an interpretation key which compared the thermal image to a simultaneously acquired panchromatic photograph. Simplified, the key used to classify hot areas portrayed on the thermal image was as follows: (1) hot springs and pools were readily visible and identified on the aerial photo. (2) mud pots and gas vents or fumaroles were generally surrounded by deposits, and in many cases cones, of white kaolinite clay or sinter (silicon dioxide). These deposits appear quite white on aerial photography. Occasionally fumaroles are not surrounded by such deposits but can be readily identified by their "point source" nature on the thermal image. (3) steaming ground, on the other hand, may or may not be evidenced by a white surface deposit on the panchromatic photos. Very hot steaming ground is accompanied by hydrothermal alteration of the surface material. Generally the paucity of vegetation on the warmest portion of moderately hot steaming ground exposes some of the soil surface which appears grey or black on the aerial photo. Subtly heated steaming ground is covered with vegetation which is readily identified on newer infrared ektachrome photos.

Using this interpretation key the majority of the hot areas portrayed on the thermal images were rejected as probable direct, surface hydrothermal manifestations. Next, the hot areas identified as steaming or hot ground were checked against a topographic map to determine if they were reasonably accessible for study and had been, and would continue to remain, reasonably free of human disturbance. The available imagery of sufficient quality, which covered roughly 2% of the Park, was used to locate approximately 12 sites. These remaining steaming ground areas were rechecked on the aerial photographs to eliminate those which bordered upon poorly drained areas which would inhibit the growth of lodgepole pine and to identify those which occurred in or bordered upon vegetated areas, especially tree covered areas. Finally, the areas were field-checked and two sites selected for detailed study.

DESCRIPTION OF RUSH LAKE SITE

The Rush Lake area is in the Lower Geyser Basin, 0.6 miles from Fountain Freight Road. It is seldom visited. It consists of several small ridges formed by glacial and hydrothermal agents adjacent to the Firehole River and is almost surrounded by a large, poorly drained area called Fountain Flats. Inspection of the thermal image of this area, together with the panchromatic photograph, showed that several areas met the requirements for steaming ground ("A", Fig. 1).

The large, hot, irregularly shaped areas along both sides of the meadow on the thermal image are almost undetectable on the stereogram. The right area is bordered on the southeast side by an even-aged, closed stand of lodgepole pine. This growth form of lodgepole pine is identifiable by its fine texture and high density on the stereogram and by the knowledge that lodgepole pine is the dominant tree species in the lower elevations of the Park.

Summer field inspection of this area resulted in a sketch map of the site (Fig. 2). Almost completely surrounding the meadow are areas of steaming ground. All these areas showed evidence of the expanding nature of the steaming ground. The most dramatic evidence of the dynamic nature of the steaming ground occurred along the south and east of the meadow where the expanding, heated area was encroaching upon an approximately 47-year-old, ever-aged, closed lodgepole pine stand. The original thermal activity occurred at the border between the pine stand and the meadow and has been steadily expanding in all directions since that time. The expanding zones of thermal flux are clearly evident in the present condition of the trees.

The trees near the original border between the lodgepole stand and the meadow were all dead and had been felled by the wind. Further east was a zone of standing dead trees without foliage which had not yet been downed by the wind. Adjacent to this band the stand was a band of dying lodgepole pines whose foliage was present but was rust colored and dry, indicating that their heat tolerance was exceeded no more than 2 years previously. At a greater distance the stand was essentially normal. There were undoubtedly additional zones occurring in this area where the functioning pines were physiologically affected by smaller increases in earth heat flux.

The distribution of the understory vegetation of the several lodgepole pine zones of the Rush Lake Site indicated additional effects of the existing thermal gradients. The average temperature of the ground at 10 cm depth in the hottest area of bare ground was 72°C on a cloudy day in August of 1965. Soil temperatures were measured with a contact pyrometer at the bottom of a 10 cm hole immediately after it was dug. A moss-lichen community occurred in the zone of downed lodgepole. The average temperature in this zone at 10 cm depth was 58°C. The most complex area of ground vegetation existed beneath the standing dead trees where the ground temperature was 26°C. Approximately 30 species occur in this area and almost completely cover the ground surface. A third distinct area of ground vegetation existed under the dying lodgepole where the average ground temperature was 18°C. As would be expected there was almost no ground vegetation under the closed lodgepole pine canopy due primarily to lack of light. Here the average ground tempera-

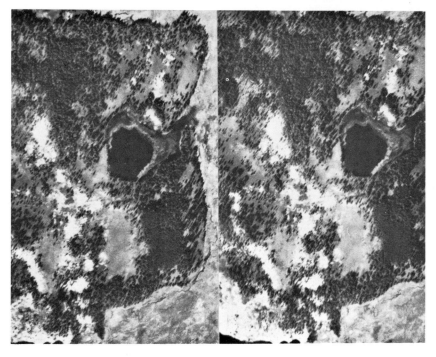

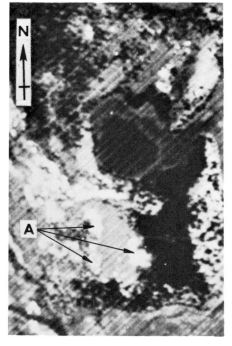

Figure 1. A comparison of a stereogram of panchromatic aerial film and a thermal infrared image of the Rush Lake Area of Yellowstone National Park. All images were acquired at 1000 hrs. on May 22, 1961. (Courtesy of U. S. Army Cold Regions Research and Engineering Laboratory.)

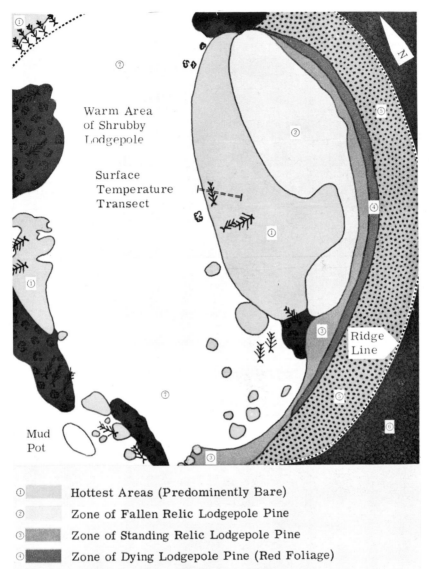

Warm Area
of Shrubby
Lodgepole

Surface
Temperature
Transect

Ridge
Line

Mud
Pot

① Hottest Areas (Predominently Bare)
② Zone of Fallen Relic Lodgepole Pine
③ Zone of Standing Relic Lodgepole Pine
④ Zone of Dying Lodgepole Pine (Red Foliage)
⑤ Closed Stand of Lodgepole Pine (47 Years Old)
⑥ Uneven Aged Lodgepole Pine
⑦ Meadow

Figure 2. Sketch map of the areas of hot earth surrounding a meadow in the Rush
    Lake Area prepared on Aug. 21, 1965. The dynamic nature of one of the hot
    areas can be seen from its effect on the adjacent lodgepole pine stand.

ture was 15°C and the air temperature 2 cm above the ground was
17°C. All these variations in the understory vegetation cannot necessarily be attributed directly to differences in the thermal flux. Many of
the changes in understory vegetation are controlled by variations in
other physical factors such as light and moisture, coupled with alteration
of the lodgepole pine cover by the increasing thermal flux.

A historical comparison of aerial photographs of the site (Fig. 3)
gives dramatic evidence of its dynamic nature. In 1954 the lodgepole
stand was intact (Fig. 3, upper left). In 1961 only a few individual

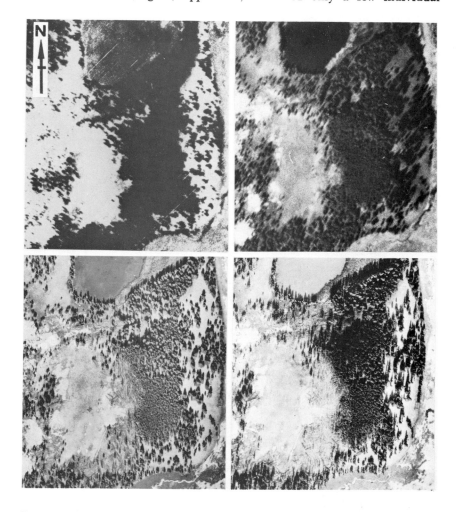

Figure 3. A comparison of aerial photographs of the Rush Lake Site showing the
changes in the lodgepole pine stand over a 13 year span. The upper left photo
(Sept. 21, 1954) and the upper right photo (May 22, 1961) show little change
in the tree stand. The lower left photo (Aug. 31, 1965) and lower right photo
(Sept. 19, 1957) show considerable and progressive alteration of the stand.

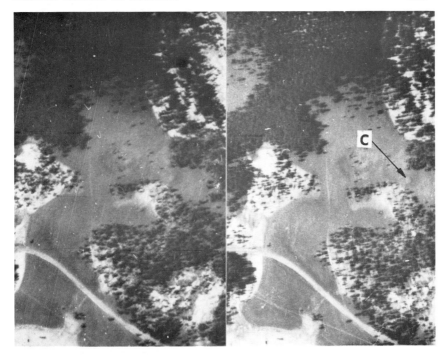

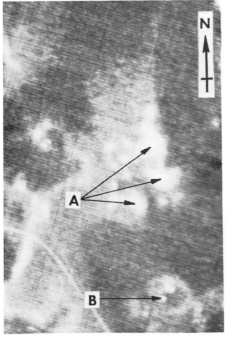

Figure 4. A comparison of a stereogram of panchromatic aerial film and a thermal infrared image of the Black Sand Area of Yellowstone National Park. The panchromatic photos were acquired at 1000 hrs. on May 22, 1961 and the thermal image at 0645 hrs. on May 20, 1961. (Courtesy of U. S. Army Cold Regions Research and Engineering Laboratory.)

trees had died (Fig. 3, upper right). By 1965 the zone of downed trees had expanded considerably (Fig 3, lower left) and by 1967 it included approximately ⅓ of the original stand (Fig. 3, lower right). This comparison of old aerial photographs demonstrates that the beginning of the major intrusion of steaming ground into this lodgepole pine stand apparently coincided with a major earthquake in the Yellowstone region in 1959. The 1959 earthquake also strongly affected other types of hydrothermal features in the immediate area (Fischer 1960). Six hundred trees have been sampled by an increment borer on a block transect 10 meters wide and 200 meters long parallel to the gradient in soil temperature and crossing all the lodgepole pine zones. Identification of the year of cessation of growth from the ring structure of each tree sampled as a function of its ground position is expected to pinpoint the year of the beginning of the expansion of the steaming ground. Similarly, the rate of decrease or increase in the expansion of the site will also be indicated. Measurement and analysis of the widths of the growth rings for each tree sampled may identify pre-death physiological effects on the growth of each tree as the high soil temperature zones expanded.

DESCRIPTION OF THE BLACK SAND SITE

Less than ½ mile from Old Faithful within the Black Sand Basin is a different type of anomalously hot earth. The site (Fig. 4 "A") borders an uneven-aged stand of lodgepole pine. It meets the required interpretation criteria: (1) it is hot on the infrared image with little or no indication on the panchromatic photo, (2) there is no sign of a hot runoff stream in the area in either image, (3) there is no evidence of a very hot point source from a gas vent or hot pool (e.g., Fig. 4 "B"), (4) the thermal area borders directly upon the lodgepole stand and is on well drained, high ground, and (5) the boundaries of the hot areas are rather indistinct.

Winter field inspections of a portion of the area (center arrow at "A", Fig. 4) resulted in a sketch map (Fig. 5). The hottest portion of this site was expanding selectively in one direction along the south edge of the snow free area as indicated by the presence of dying lodgepole pine with reddish-brown foliage. Annual growth ring samples have been taken from 75 trees in this area for future analysis. It should be possible to determine the year of the beginning of this expansion by tree ring dating of the relics.

At the center of the snow free area is a ring of black, crushed obsidian loosely piled, slightly higher than the inside or outside area. The inside of this ring is almost a perfect circle 11 m in diameter (Fig. 4 "C") This enclosed circular area is one of the hottest parts of the site. It contains vegetation not found outside the ring in the remainder of the snow free area.

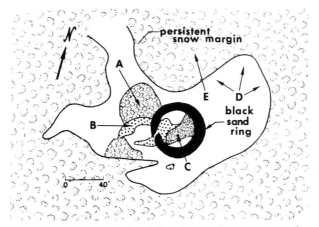

Figure 5. Sketch map of the Black Sand Site hot earth area prepared on Jan. 23, 1966. Five different areas of vegetation, "A" to "E", are shown.

Five distinct types of ground vegetation existed on the site (Fig. 5). Four of these vegetation types occurred within the area which remained snow free during the winter ("A" to "D", Fig. 5) when the normal snow accumulation on the ground elsewhere reaches 75 to 125 cm. The vegetation of the three warmest of these areas ("A" to "C", Fig. 5) is highly specialized, consisting of a single species each. One of these occurring on the hottest ground consists of tiny rosettes of *Euphorbia glyptosperma* which can be found in full flower under the extreme conditions present during January. The third area contains *Panicum thermale* which is somewhat depauperate during winter. *P. thermale* occurs exclusively on anomalously warm, acid ground (Hitchcock and Chase 1910) and was originally collected and identified around hot springs in the Shasta area of California (Brewer 1868). The functioning vegetation in all three areas has assumed a form which is restricted to 1 or 2 cm in height. The first 1 or 2 cm of air above the surface of the hot earth, the laminar air layer, remains warm enough to permit growth of green vegetation throughout the winter.

GEOPHYSICS OF THE BLACK SAND SITE

A detailed investigation of the energy relationships of the Black Sand Site was undertaken to determine how steaming ground manifests itself in the surface layers of the soil.

Snow Patterns

The depth of snow around steaming ground areas has been used as an indication of the distribution of soil temperature and heat flux (Dort 1965, 1968; Miller 1967; Lindl, 1968). The procedure by which these measurements are translated into meaningful soil temperature or heat flux maps has not yet been completely solved. Maps of differential

snow melt around sites of steaming ground do yield quick information
on the qualitative spatial distribution of soil heat flux over the site.
For example, a single snow depth contour corresponds approximately
to an isotherm of the soil surface and an isogram of the net outward
energy losses from the soil.

A snow depth map for the site was prepared in 1967 which indicates
the distribution of soil heat flux outside of the snow free margin (Fig.
6, upper). This map shows that the snow depth gradient around the
snow-free portion of the site was initially quite steep but became shallower
with increasing distance from the snow margin. As the snow margin
retreated the newly exposed soil was 'recharged' with energy since it
was no longer supplying the latent heat of fusion to melt snow nor was
the cold, 0°C melt water percolating into the soil. The temperature
profile for an area of soil freed of snow cover gradually shifted to a new,
higher steady state representing a net gain in the storage of energy in

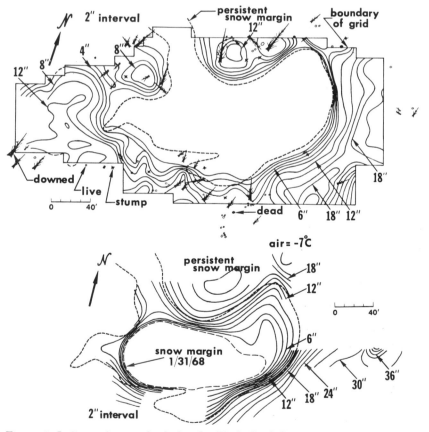

Figure 6. Isolines of snow depth for the Black Sand Site qualitatively illustrating
the distribution of the anomalous soil energy flux over the site. Upper: Jan. 4,
1967. Lower: Jan. 31, 1968.

the soil profile. This extra stored energy was available to accelerate the melting of new snow falling on this area and was one of the factors resulting in the steep gradients found at the edge of the snow. Only just after a heavy snowfall was any measurable depth of snow found within the area surrounded by the "persistent snow margin" (Fig 6, lower). At the central portion of the site a heavy snowfall of 40 to 50 cm in the course of a day will not remain on the ground. Snow depths are therefore not useful to indicate the energy flow in the hottest portion of steaming ground.

Maps of the margins of the snow melting off the site during and after heavy snowstorms are a qualitative measure of energy flux distribution over the central portion of the site. A map of the snow margins was maintained for an eleven-day period in January of 1968 (Fig. 7). Heavy

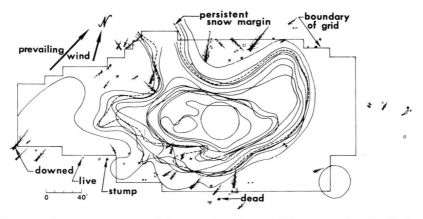

Figure 7. Snow margins mapped between Jan. 24 and Feb. 5, 1968 for the Black Sand Site. These snow margins illustrate the distribution of the anomalous soil energy flux over the central, hottest portion of the site where it cannot be observed in terms of snow depths.

snows fell during this period and several margins were mapped just after or during these snow falls. Each snow margin corresponded approximately to an isogram of energy lost through the soil on the hottest portion of the site which cannot be determined from depth profiles of the persistent snow cover. These measurements illustrate the general pattern of the net energy dissipated by the central portion of the area. The innermost snow margin mapped during a heavy snowstorm showed two localized hot spots at the center of the site. These corresponded with the vegetation distribution (Fig. 5) and the soil temperatures mapped the previous year. The outermost margin recorded shows a small subtle circular warm spot to the southeast which was exposed only after 10 days free of snowfall. These snow margins also illustrate that the soil heat flux gradients were not radially symmetrical. The margins retreated in some directions (e.g., to the NNE, NNW, and W) much more rapidly than in others. One of these 1968 snow margins corresponded almost exactly to the 1967 snow margin suggesting that there had been no differential change in the distribution of the energy flux over the site in the intervening year. This is noteworthy as the energy flow pattern in the previous few years had been moving northward in the NNE corner of the area as evidenced by several dead and dying trees at that point. No evidence of new advance could be detected in the vegetation in 1968.

Soil Profile Conditions

A winter soil moisture profile at the hottest point on the site delineates a zone of very high soil water content at 64 cm inches depth compared with the soil moisture profile taken at the snow margin (Fig. 8, left). The water content of the soil at the center of the site increased from 10% of the dry weight of the soil to 100% at 64 cm inches and then decreased markedly with increasing depth. In contrast, the soil moisture at the snow margin is a constant 10% throughout the profile.

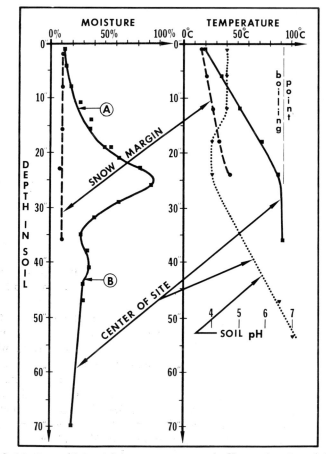

Figure 8. Soil moisture (by weight), temperature, and pH as a function of depth for the center and snow margin of the Black Sand Site (Jan. 1968).

It will be shown that little mass transfer of water vapor takes place above the top of the water layer. Condensation of water vapor occurs at depths appropriate for conductive transfer of heat resulting from the condensation. Most but not all of the water vapor condenses near 25 inches depth giving high soil moisture at that depth. Below the depth of maximum water content the soil temperature is at the boiling point of water while at successively shallower depths it decreases to a much lower surface temperature (Fig. 8, right).

The soil profile has been extensively altered within the zone of the steam condensation. Hydrogen sulfide and carbon dioxide are present in the steam in small but significant amounts. Which of these acids plays the more important part in the minerological processes is currently under investigation. The H_2S combines with oxygen and water in the upper meter of the soil to form sulfuric acid. This process is probably aided by sulfur oxidizing bacteria (Kaplan 1955; Schoen 1967). The decreasing temperatures encountered in the high soil moisture zone allow the CO_2 to go into solution in the soil water and form carbonic acid. The pH of the upper portion of the soil is between 4 and 5 and increases to neutral as one reaches the depth of live steam (Fig. 8, right). The low-pH condensate and high soil temperature have altered the parent soil material, which in this area is a water laid sand and gravel. Considerable zonation occurs in the alteration products. Below the present water layer in the steam zone the sand and gravel have been reduced to a resistant skeletal structure. This material is quite light in weight and like styrofoam in structure. Above and in the present water layer are brightly colored bands of clays and silts produced in place by the hot condensate. The most prominent material is kaolinite (Muffler 1968). At the center of the site at depths of the maximum soil moisture content the greatest percentage of clay and silt are present in the profile (Fig. 9, left). These silts and clays have higher water retention capabilities, when compared with the sand making up the remainder of the profile. A bulk density of 1.6 in the first 25 cm of the soil at the center of the site approximately equals that at the snow margin (Fig. 9, right). Below 25 cm the bulk density at the center of the site is one-half of that at the snow margin, indicating that many slowly soluble constituents are flushed from the soil profile by the downward percolating condensate.

Soil Conditions On The Site

The distribution of the soil moisture anomaly was mapped on a 3 m grid along with the measurement of soil temperature profiles. As the moisture content of the soil increased a depth was reached at which a 0.5 cm soil auger could be forced further into the soil without turning. This was defined as the "top" of the high-water layer ("A", Fig. 8). The depth at which the auger could no longer be forced into soil without turning was considered the "bottom" ("B", Fig. 8). These points were found to occur at moisture contents of 20 to 30%. The top of the high-water layer, defined in this manner, varies between 25 cm and 60 cm below

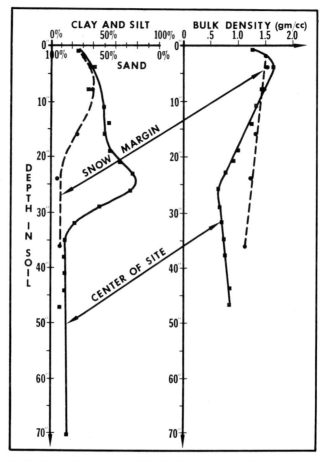

Figure 9. Soil particle size and bulk density as a function of depth for the center and snow margin of the Black Sand Site (Jan. 1968).

the soil surface and the thickness ranges between 25 cm and 150 cm (Fig. 10, lower). A major portion of the steam rises in the center of the site where it condenses near the surface. Less steam occurs away from the center of the site and less latent heat of vaporization must be conducted away, lower temperature gradients are required for conduction of this heat, therefore steam condenses deeper in the soil. The water formed in this condensation process presumably percolates downward from the soil water layer in a two-phase, steam-water system and rejoins the ground water table. This system appears to resemble a large convective cell within the soil.

Measurements of the moisture content of the surface 8 cm of soil in the summer also indicated the water layer (Fig. 11). The center of the steaming ground area had a higher surface moisture content in the summer than the immediately surrounding area due to the con-

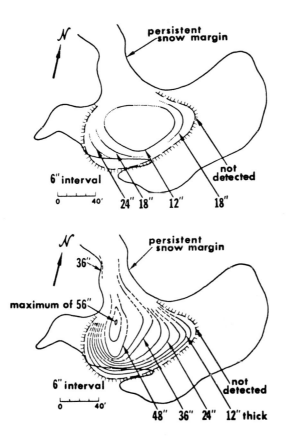

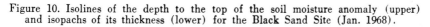

Figure 10. Isolines of the depth to the top of the soil moisture anomaly (upper) and isopachs of its thickness (lower) for the Black Sand Site (Jan. 1968).

densate. If an artificial source of soil moisture were not present, the center and hottest portion of the site would be the driest due to the corresponding increase in surface evaporation due to its higher surface temperature. The water layer is shallower in the summer than winter as a result of the increased air temperature and drier soils yielding lower values for soil thermal conductivity. Therefore the soil might not conduct energy out at a sufficient rate to liberate the latent heat of vaporization of all the steady state mass of steam supplied. A large amount of the energy flow from the site in the summer could be in the form of mass transfer of water vapor. This fluctuation in the depth of the water layer 'across' the soil surface may account for the "Indian Ring" of obsidian gravel previously described. The water layer could occur "at" the surface of the soil on a warm spring day. If nighttime temperature drops considerably below freezing the water at the surface would freeze and expand radially. Operating over several

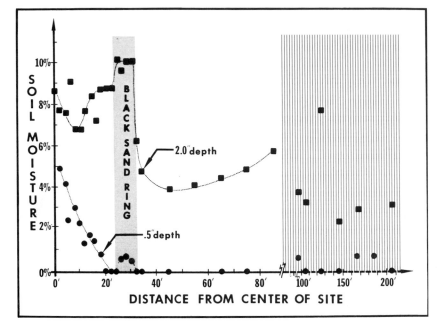

Figure 11. A transect of the moisture (by weight) at the 1″ and 3″ depth in
the soil as a function of the distance from the center of the Black Sand Site.
At both depths the highest soil moisture is found in the central, hottest portion
of the site. (Sept. 22, 1967).

hundred years this mechanism might have transported and piled the
soil surface into the 11 m ring from which the smaller soil particles
have been differentially sorted or subsequently washed by rainfall leaving
only the coarser obsidian gravel.

Upward movement of some precipitates also takes place through
the soil. The hotter the soil, the harder and more cemented the surface (Fig. 12). A certain amount of solute appears to have been
translocated upward by capillary action to offset the high evaporation at
the surface. The residue precipitates have cemented the first few centimeters of the soil profile. The first 2 cm of material covering the steaming
ground areas was loose gravel but the next 2 to 5 cm have been cemented
in proportion to the steepness of the slope of soil temperature profile. At
the hottest positions on the sites the surface had been cemented to a rock-
like consistency which could only be penetrated by an auger with
considerable effort.

Energy Relations In The Soil

Winter soil temperature profiles on the Black Sand Site were essentially linear and controlled by simple heat conduction mechanisms
above the condensed water layer (Fig. 13). In the surface layers of the
soil in the winter there was no upward mass transfer of water vapor.

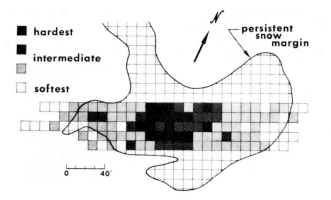

Figure 12. Distribution of the soil surface hardness for the Black Sand Site
showing the correlation between the most cemented surface area and the central,
hottest portion of the site (Jan. 1968).

The energy given up by the condensing steam at 64 cm or more
is much greater than day to day changes in the energy flux into and
out of the soil. Changes in the temperature profile due to short term
meteorological and diurnal fluctuations in the surface energy budget
are minor at 15 cm although they are important at 2.5 cm.

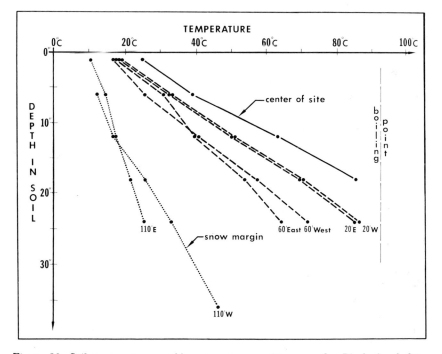

Figure 13. Soil temperature profiles at various positions on the Black Sand Site.
These curves for the soil above the moisture anomaly are essentially linear (Jan.
1968).

Extensive measurements of soil temperature as a function of depth were made over the site for depths up to 90 cm. Isogeothermal maps of the temperature at various depths (Fig. 14) correlated well with the presence of the soil water layer (Fig. 9). The simple conduction equation can be used to describe the net loss of energy from this steaming ground area. The net energy loss at any point is therefore:

$$q = \frac{k\ T_2\text{-}T_1}{x}$$

where:　　q = rate of heat loss in cal/cm²-sec

k = thermal conductivity = $\dfrac{3 \times 10^{-3}\ cal}{°C\text{-}cm\text{-}sec}$

T_2 = temperature in °C at 18″

T_1 = temperature in °C at 6″

x = 12″ = 30.5cm

The temperature gradient between 15 and 45 cm was found to be essentially linear over the site (Fig. 13) resulting in the values selected for T_2, T_1, and x. Thermal conductivity (k) ranges from .001 to .008 cal/°C-cm-sec for various soil types and moisture contents (Ingersoll and Zobel 1954). Further processing of the soil sample information will allow a value to be computed for k which will vary over the site. For the initial computation k has been assumed to be .003 cal/°C-cm-sec.

The heat loss from each 3 m grid square for which temperature profile data was available was computed and plotted (Fig. 15, upper). The heat loss from the hottest portion of the site was between .0045 and .0060 cal/cm²-sec. At the persistent snow margin it was .0015 to .0030 cal/cm²-sec These values are greater than a normal representative terrestrial heat flux of 1.5 micro-cal/cm²-sec (Lee 1965) by 3000 to 4000 and 1000 to 2000 times, respectively. The computed heat flow rates were extrapolated to the remainder of the area (Fig. 15, lower) in terms of the snow measurements (Figs. 6 and 7). Integrating the heat flow over the control site for the area inside the persistent snow margin gave a total heat loss of 55 x 10³cal/sec.

NEW INFRARED IMAGERY OF THE BLACK SAND SITE

New thermal infrared scanner imagery and aerial photographs of the Black Sand Area were taken during the summer of 1967 for the U. S. Geological Survey by the University of Michigan (Fig. 16). The Black Sand Site studied in detail as a control area is indicated on new thermal images whose temperature range was 0° to 20° C (Fig. 16, "A"). The information collected on the control site can be used to extend meaningful interpretation to the remainder of the image.

The surface of the soil in the steaming ground areas studied was glacial outwash gravels dominated by black obsidian gravel and measurements of its emissivity have shown that it closely approximates a black body radiator. The apparent surface temperatures indicated in the thermal image are therefore good approximations of the real surface

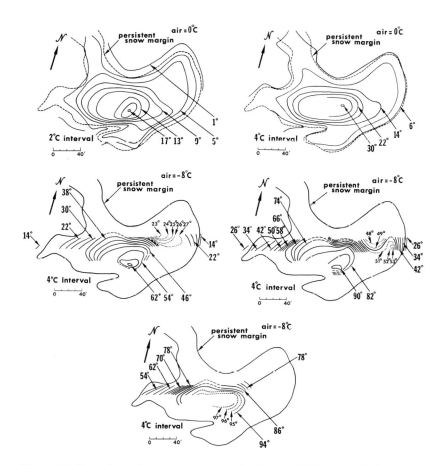

Figure 14. Isogeothermal maps for various depths on the Black Sand Site showing
the spatial distribution of soil temperature. Upper left: 1 inch depth. Upper
right: 4 inch depth. Middle left: 1 foot depth. Middle right: 2 foot depth.
Lower: 3 foot depth. (Feb. 3 and 5, 1967).

temperatures of the soil. A map of the soil surface temperature was
made on the ground with a radiometer (Fig 17) at nearly the same
time as the imagery was obtained. This surface temperature map
correlates closely with the snow depth, energy flux, and subsurface soil
temperature maps illustrated previously as well as with the thermal
image. The small warm spot to the SE of the control site (Fig. 7)
can just be detected on the image (Fig. 16, "B"). Ground radiometer
measurements of the apparent surface temperature of this area taken
simultaneously with the image indicated that it is approximately 3 to
4°C warmer than the background areas surrounding it.

The control area is readily seen to be part of a larger complex
of steaming ground sites manifesting similar ranges of surface tempera-

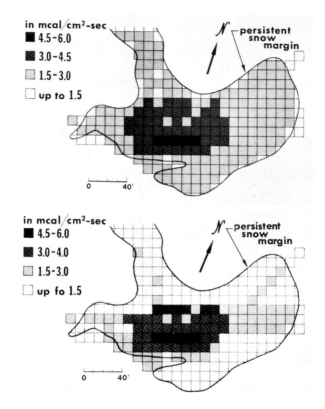

Figure 15. Net outward energy flux from the Black Sand Site. The upper map
is the computed energy loss for 10′ grid squares using the simple heat conduction
equation. The lower map is the extrapolation of these values to all the
area inside the persistent snow margin based on the snow depth and snow
margin maps.

tures. These sites collectively were a roughly circular formation (Fig. 16,
"C"). This formation of steaming ground is one of several such which
occur in the Black Sand Area. The persistent snow margin and the 4°C
surface temperature contour measured on the ground (Fig. 17) correspond roughly to the margin of the heated area of the control site
identifiable on the thermal imagery and found to liberate 55 x 10³cal/sec.
A second steaming ground area (Fig. 16, "D") in this formation resembles the control site in area and surface temperature and is assumed
to lose an equal 55 x 10³cal/sec. The remainder of the steaming ground
in the circular formation was lower in temperature but greater in area
than these two sites and was estimated to contribute 60 x 10³cal/sec.
The heat loss from the total formation is thus approximately 170 x 10³cal/
sec or 170 Kcal/sec. This amount of heat would result from condensing
approximately 280 gm/sec of steam to water.

The thermal image shows what appears to be a 300 m subsurface

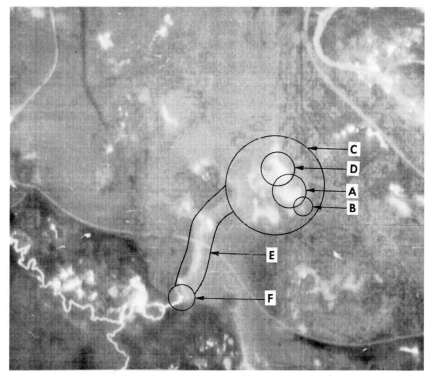

Figure 16. Thermal infrared image acquired at 0300 hrs. on Sept. 20, 1967 of the Black Sand Site and surroundings. The temperature range represented by the black to white tones in this image is 0 to 20°C, respectively, with the white tone representing everything equal to or greater than 20°C. (Courtesy of U. S. Geological Survey.)

path (Fig. 16, "E") of hot water from the circular formation to a point at the intersection of the ground water table and topographic surface shown on the stereogram (Fig. 4) where it manifests itself as a hot spring (Fig. 16, "F"). The only winter evidence of this path at the surface of the ground was the absence or shallower depth of snow. The February temperature of the hot spring was found to vary between 70 and 70.5°C over the 2-week measurement period. This temperature range would be commensurate with the cooling of the 92.5°C condensate as it flowed from the formation to the hot spring. The energy given up by cooling the water from 92.5°C to 70.5°C would heat the ground above the seepage channel as indicated. Using a "V" notch weir the flow of the spring over the same 2 week period was measured as a uniform .12 cubic feet per second This is 3400 gm/sec of water which is 12 times more water than could be produced by the steaming ground formation. The U. S. Geological Survey has also found that this spring has a high mineral load, especially chloride, which proves that this water is not a simple condensate of steam.

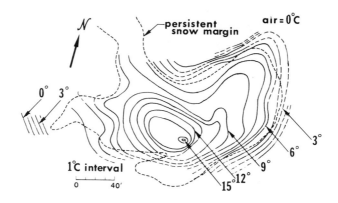

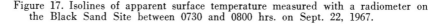

Figure 17. Isolines of apparent surface temperature measured with a radiometer on the Black Sand Site between 0730 and 0800 hrs. on Sept. 22, 1967.

SUMMARY

This study was originally designed as a field program to determine the application of remote multispectral sensing techniques to obtaining synoptic information of ecological significance. Indeed, the study sites were located using remote sensing techniques, more specifically an inter-comparison of thermal and visible band imagery. The unusual nature of the sites has however necessitated an extensive and continuing study of the energy and biological relationships existing in the test area. While these ground investigations are far from complete a basic under-standing of the mechanisms controlling the ecology of the sites has begun to emerge. With this limited knowledge of the ground conditions it has been possible to demonstrate the potential usefulness of thermal imagery to extrapolate information collected on control sites to larger areas. While not complete, processing of the available simultaneous multispectral imagery in numerous other spectral bands acquired for these same sites also promises fruitful results in such areas as identifica-tion and mapping of ground surface material, mapping of percent vegetative cover as a function of soil surface temperature, and mapping of the various physiological zones occurring in the vegetation, more specifically, lodgepole pine.

REFERENCES

Benseman, R. F. 1957. Changes in the heat output from steaming ground at Orakei Korako and Wairakei. N.Z. Dept. Science and Industrial Res. Dominion Physical Lab, Rept. No. R277.

Benseman, R. F. 1959a. The calorimetry of steaming ground in thermal areas. J. Geophys. Res. 64:123-26.

Benseman, R. F. 1959b. Estimating the total heat output of natural thermal regions. J. Geophys. Res. 64:1057-62.

Brewer, W. H. 1968. Notice of plants found growing in hot springs in California. Proc. Calif. Acad. Natl. Sci. 1863-67, No. 3.

Brock, T. D. 1967a. Relationship between standing crop and primary productivity along a hot spring thermal gradient. Ecology 48:566-71.

Brock, T. D. 1967b. Life at high temperatures. Science 158:1012-19.

Dort, W., Jr. 1965. Yellowstone Field Research Expedition V. Atmospheric Sciences Res. Center, State Univ. of N. Y., Albany, pp. 56-61.

Dort, W., Jr. 1968. Yellowstone Field Research Expedition VIII. Atmospheric Sciences Res. Center, State Univ. of N. Y., Albany.

Elder, J. W. 1965. Physical processes in geothermal areas. In W. H. K. Lee, Terrestrial heat flow. Am. Geophys. Union, Publ. No. 1288, cp. 8.

Fischer, W. A. 1960. Yellowstone's living geology. Yellowstone Library and Museum Assoc., Yellowstone National Park, Wyoming.

Hitchcock, A. S and A. Chase. 1910. The North American species of Panicum. Contr. U. S. Natl. Herbarium 15:231-2.

Ingersoll, L. R. and O. J. Zobel. 1954. Heat conduction with engineering, geological, and other applications. Univ. Wisconsin Press, Madison.

Kaplan, I. R. 1955. Evidence of microbiological activity in some of the geothermal regions of New Zealand. N.Z.J. Science & Technology 37B:650-62.

Kintzinger, P. R. 1956. Geothermal survey of hot ground near Lordsburg, New Mexico. Science 124:629-30.

Lee, W. H. K. 1963. Heat flow data analysis. Rev. Geophys. 1:449-79.

Lindl, J. 1968. Yellowstone Field Research Expedition VIII. Atmospheric Sciences Res. Center, State Univ. of N. Y, Albany.

Marler, G. D. 1964. Effects of the Hebgen Lake earthquake of August 17, 1959 on the hot springs of the Firehole Geyser Basins Yellowstone National Park. U. S. Geol. Surv. Prof. Paper 435-Q:185-97.

McLerran, J. H. 1965. Thermal mapping of Yellowstone National Park. Proc. 3rd Symp. Remote Sensing Environment. Univ. Michigan, Ann Arbor. pp. 517-30.

Miller, L. D. 1966. Location of anomalously hot earth with infrared imagery in Yellowstone National Park. Proc. 4th Symp. Remote Sensing of Environment. Univ. Michigan, Ann Arbor. pp. 751-69.

Miller, L. D. 1967. Winter observations on steaming ground in Yellowstone National Park, 1967. Yellowstone Field Research Expedition VII, Atmospheric Sciences Res. Center, State Univ. of N. Y., Albany, pp. 112-120.

Muffler, L. J. P. 1968. Personal communication. U. S. Geol. Survey, Menlo Park, California.

Schoen, R. and G. C. Ehrlich. 1967. Bacterial origin of sulfuric acid in sulfurous hot springs. U. S. Geol. Survey, Menlo Park, Calif.

8

Observations on Interpretation of Vegetation from Infrared Imagery

DAVID K. WEAVER
WILLIAM E. BUTLER
CHARLES E. OLSON, JR.

INTRODUCTION

RECENT reports on interpretation of vegetation from thermal infrared imagery seem contradictory. In 1967, Olson indicated that infrared imagery obtained near mid-day was more desirable for identification of crop types than night imagery. However, Stingelin (1968) reported better results with night imagery when working in marsh and forest conditions. In both studies the imagery was obtained with filtered indium antimonide detectors; Olson's was filtered to the 4.5 to 5.5µ, and Stingelin's to the 3.0 to 5.0µ wavelength bands.

During 1966, and again in 1967, ground data and airborne infrared imagery were obtained concurrently for several areas near Ann Arbor, Michigan. The imagery was supplied by the Bendix Aerospace Division of the Bendix Corporation as contributions to two University of Michigan/National Science Foundation short courses on remote sensing. Ground data were gathered by thirty college-teacher participants in each of the two short courses. Subsequent comparison of participant field reports with the infrared imagery produced results which support the conclusions of both Olson and Stingelin.

IMAGE PARAMETERS

Imagery of the test areas was obtained in mid-afternoon and at approximately 11 p.m. on June 24, 1966, and on June 20 and 23, 1967. A Bendix Thermomapper with an indium antimonide detector filtered to the 3.7 to 5.5 micron band was used in each case. Weather conditions preceding the three flight periods varied.

The spring of 1966 was dry, with no precipitation recorded at Ann Arbor from May 12 (1.68 inches) to June 9 (0.61 inch). Thunder

132

storm activity between June 13 and 17 produced 1.77 inches of rain at the Ann Arbor station and unknown amounts at the test sites ten miles to the west and northwest. No additional precipitation was observed until June 27. Air temperatures exceeded 27°C on each of the seven days preceding the 1966 flights, and exceeded 32°C on each of the last three days. A reading of 34.5°C was recorded in Ann Arbor on June 24.

In 1967, May was extremely dry in Southeastern Michigan, with only 1.00 inch of rain recorded in Ann Arbor. Of this, 0.51 inch was reported on May 10. The next rains of 0.1 inch or more occurred on June 7 and 8 when a total of 1.61 inches was recorded. An additional 1.13 inches fell on June 16 and 17. On June 21 a sharp evening shower followed by a slow drizzle brought 0.48 inch of precipitation. Air temperatures reached the 30°C's on June 14 and 15, but peaked between 25° and 28°C from June 17 through June 23.

The differences in weather conditions described above, coupled with day and night flight scheduling, provided imagery with interesting tonal variations. Because infrared scanners are AC-coupled, tone comparisons between imagery from different flights have no quantitative significance. Relative tone values — or tonal rankings — within single strips of imagery are more useful. To facilitate tonal ranking within image strips, an arbitrary gray scale was assembled from commercially available construction papers and provided a five-step scale with reasonably uniform density changes between steps. This gray scale was used to rank, to the nearest half-step, the tonal value of each cover type shown in the imagery. Ground data collected at the time of the overflight confirm that light tones correspond to higher radiometric temperatures than dark tones.

RESULTS

The range of tonal values was greater in the daytime imagery than in that obtained at night. Tone rankings for the June 20, 1967, imagery were typical, and are presented in Table 1. The day/night tone reversal characteristic of water areas was apparent in both the imagery (Figure 1) and the tonal values. In the daytime imagery different cultivated crops appeared in detectably different tones, with deep-rooted alfalfa much darker than shallow-rooted corn, except in one case in which the alfalfa had recently been cut for hay. Swamp areas, and stands of coniferous or broadleaved trees, were as dark toned as alfalfa, and these species could not be separated on the basis of tone.

In the night imagery the tone of agricultural crops was much less variable and separation of crop types was difficult or impossible. However, natural vegetation showed greater tonal differentiation at night than in the daytime. Swampy areas were still quite dark but stands of trees were lighter toned, with plantations of pine (*Pinus* spp.) somewhat lighter toned than broadleaved stands. Similar results were ob-

Table 1. Comparison of Tonal Differences in Day and Night Infrared Imagery in the 3.7 to 5.5 Micron Wavelength Band; June 20, 1967, Washtenaw Co., Michigan.

Tone Value	Day	Night
0.5 (white)	Roof tops	—
1.0	Asphalt road	Asphalt road · Open water
1.5	Gravel road Concrete	Gravel road Concrete
2.0	Lawn grass	Red pine White pine Corn
2.5	Corn Pasture	Lawn grass Broadleaved trees
3.0	Hay	Hay Lawn grass (watered, wet) Pasture Alfalfa
3.5	—	Roof tops (indistinguishable or slightly darker than surroundings) Swamp
4.0	Swamp Alfalfa Red pine White pine Broadleaved trees	—
4.5		—
5.0 (black)	Open water	—

tained in other areas where, as in the case just described, the terrain is relatively low and most of the higher ground had been cleared for agriculture.

In all cases in which our analyses dealt with bottomland stands, swamps, or conditions where soil moisture was not in short supply, night imagery provided better differentiation between natural vegetation types than daytime imagery. This is in agreement with the results of Stingelin (1968). Similar results were not obtained on dryer sites.

At two locations forested areas on top of glacial moraines were imaged. In both areas pine plantations are interspersed among stands of broadleaved species, primarily oak (*Quercus* spp.). Tonal differences between pine and broadleaved types were consistently observed in the daytime imagery but were not detectable at night (Fig. 2). The pine appeared lighter in tone than the oak in the imagery obtained on all three daytime flights, but the tonal difference was greatest in 1966. In 1966, the original imagery also revealed the outer 50 m of each of the upland broadleaved stands in tones intermediate between the tone at the center of the stand and the tone of nearby pine plantations. While this suggests a horizontal thermal gradient with temperature

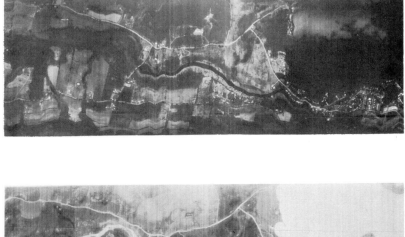

Figure 1. Day (top) and night imagery of an area near Ann Arbor, Michigan obtained on June 20, 1967, with a Bendix Thermomapper operating in the 3.7 to 5.5 micron wavelength band. Courtesy Bendix Aerospace.

decreasing toward the center of the broadleaved stands, the ground data are inadequate to confirm or deny this possibility.

Throughout our analyses we attempted to separate species in mixed stands. With the exception of the contrasts between pine and broadleaved types previously mentioned, these efforts were uniformly unsuccessful. Variations in tone within species were often greater than between species within single stands.

DISCUSSION

Moisture stress is believed to be a major factor affecting the tonal differences between plant species, or species groups, shown in infrared imagery. Under most conditions plant moisture stress shows a diurnal cycle with highest stress during early afternoon. Stress decreases rapidly after sunset and is lowest in the early morning hours. As available soil moisture is depleted, plant moisture stresses reach higher levels during the day, and do not drop as low at night, as they would if moisture were not limiting. As moisture stress increases, radiant emission from the stressed plants also increases (Weber and Olson 1967). Increasing emitted energy tends to produce lighter tones in the imagery. When

Figure 2. Infrared image of a forest area containing pine plantations (P) and
natural broadleaved stands (Q) obtained on June 24, 1966. The outer 150 feet
of each hardwood stand is lighter-toned (warmer) than the center of the same
stand. Courtesy Bendix Aerospace.

differences in moisture stress are associated with differences in species,
the differences due to stress may provide species discrimination.

The lighter tone of the pine plantations on the morainal areas during
the daytime, compared with the adjacent broadleaved stands, is consistent with the moisture stress hypothesis. Red pine (*Pinus resinosa* Ait.)
and white pine (*P. strobus* L.) are relatively shallow rooted. These
pines develop high moisture stresses faster than oaks growing on the
same site, for oaks have deeper roots and can draw water from greater
depths.

Additional support for the water stress hypothesis was provided by
comparative interpretation of the three sets of imagery. Tonal contrast
between coniferous and broadleaved stands was greatest in 1966. The
week preceding the overflight was hot and dry. In 1967, significant
rain fell three days or less before each flight and tonal contrasts between pines and broadleaved species were not as great as in 1966.
The light-toned edge effect observed with all upland broadleaved stands
imaged in 1966, but not detected in any of the 1967 imagery, may
have been another manifestation of differences in moisture stress.

All agricultural areas shown in the imagery were located on upland
sites and showed greater tonal contrasts between crops in the daytime
than at night. The superior tone separation of agricultural crops in
the daytime imagery may have been partially due to differences in water
stress. Olson (1967) was confident that such differences did exist in
Minnesota when he observed sharp tonal differences between deep-
rooted alfalfa and shallower-rooted soybeans and corn. However, the
Minnesota imagery was flown in July while our imagery was flown
in June when crop plants were small and only partially obscured the
dry surface soil beneath them. The integrated signal from the corn
fields must have included both plant and bare soil characteristics.
Differences between crops due to plant form, cultivation practices, and
times of planting and harvest all contributed to variability in tonal
renditions of cultivated vegetation.

If daytime tonal differences observed between agricultural crops, and between pines and broadleaved trees, were in fact due to differences in moisture stress, then the apparent conflict between Olson and Stingelin cited at the beginning of this chapter may have a meaningful explanation. In upland areas where differences in moisture stress between species and species groups are common, infrared imagery obtained in early afternoon when moisture stress is high is more apt to record species differences than imagery obtained at night when moisture stresses are low. Plants growing in swampy areas, or where moisture is relatively unlimited, seldom develop levels of moisture stress high enough to appreciably affect their radiant emission. In such cases, the several factors discussed by Stingelin (1968) make nighttime imagery preferable to that obtained in daylight. Since Stingelin worked in bog and marsh conditions, and Olson in upland areas, their findings need not be viewed as contradictory.

SUMMARY

Conflicting reports on the relative merits of day and night infrared imagery for interpreting vegetation prompted detailed study of available imagery of two areas near Ann Arbor, Michigan. Day and night imagery in the 3.7 to 5.5 micron band, and simultaneous ground data obtained in mid-June of 1966 and 1967 are available for the study areas. Our analysis of these data lead us to believe that nighttime imagery provides the best separation of vegetation in swamps and other areas where the water table is close to the surface. For areas of high ground, including most agricultural areas, imagery obtained in mid-afternoon when daily water stress is greatest seems distinctly superior to nighttime imagery. The data now available are consistent with the earlier suggestion by Olson (1967) that differences in moisture stress is a major factor affecting the appearance of vegetation in thermal infrared imagery.

REFERENCES

Olson, C. E., Jr. 1967. Accuracy of land-use interpretation from infrared imagery in the 4.5 to 5.5 micron band. Annals Amer. Assoc. Geographers, 57:382-388.

Stingelin, R. W. 1968. An application of infrared remote sensing to ecological studies: Bear Meadows Bog, Pennsylvania. Proc. Fifth Symp. Remote Sensing Environment, Univ. Michigan, Ann Arbor, pp. 435-440.

Weber, F. P. and C. E. Olson, Jr. 1967. Remote sensing implications of changes in physiologic structure and function of tree seedlings under moisture stress. Annual Progress Report, Forestry Remote Sensing, Laboratory, U. S. Forest Service, Berkeley, California, p. 61.

Progress in Large Animal Census by Thermal Mapping

DALE R. McCULLOUGH
CHARLES E. OLSON, JR.
LELAND M. QUEAL

INTRODUCTION

INFRARED scanning equipment, originally developed for military purposes, is proving to have a number of applications to natural resource and ecological problems. Much wider use will certainly occur as more sophisticated equipment is declassified.

Because infrared scanners detect and record heat rather than visible light, it is apparent on logical grounds that warm-blooded animals might be distinguished from a cooler background. Thus, there is the possibility for use of this instrumentation for census of animal populations (Croon *et al.* 1968). That the airborne equipment could cover large areas over a short period of time is a further advantage.

A typical size of the instantaneous field of view (IFV) of scanners is about three milliradians (three feet across for each 1000 feet of altitude). Only a large animal can increase the average "temperature" of the IFV to a level that can be distinguished from the background adjacent to the animal. This emphasizes the fact that only large animals such as deer, elk, or domestic cattle can be detected with any efficiency with present equipment. While there is no difficulty in producing scanning equipment with a much smaller IFV, decreasing the IFV is accompanied by a loss of temperature discrimination. Because sensitivity of the equipment is approaching the theoretical limit, there is little hope that small animals such as grouse or rabbits will ever be censused from the air with infrared scanners.

A SUCCESSFUL DEMONSTRATION

While some prior testing demonstrated the capacity of infrared scanners to detect large animals, the first actual success in censusing was reported by Croon *et al.* in 1968. We tested the technique on the Edwin S. George Reserve near the University of Michigan to census a population of white-tailed deer *(Odocoileus virginianus)*. This approximately two square mile, fenced area supported a herd estimated by other methods at 101 deer as compared to an infrared count of 93 positive and 5 probable deer. While the estimate of 101 animals may not be absolutely correct, it is the result of a concentrated effort using several techniques which have been applied on the Reserve for many years. The close agreement of the infrared count and the herd estimate suggests that satisfactory results can be obtained by the infrared method if it is conducted under appropriate conditions.

APPROPRIATE WAVELENGTHS

The best range of wavelengths for animal census is 8 to 14 microns in the far infrared. Fortunately, there is an "atmospheric window" in this range where radiation passes through the atmosphere with little interference (Suits 1960). The successful George Reserve overflight was filtered to the 8 to 14μ range. Subsequently we have tested equipment filtered 3.5 to 14, and 3.5 to 5.5μ. Two separate daytime overflights were conducted, one in late November and mid-January under fair to good conditions for animal detection. The weather was clear and cold, but sunny, with minimal wind and a blanket of 3 to 4 inches of snow covered the ground. Although there were 12 deer in known locations none were detected in the imagery. Two horses in a pasture adjacent to the Reserve could be identified in the 3.5 to 14 micron imagery, but only because their presence was known. We assumed that the 3.5 to 5.5μ range was unsuccessful because of the low emission at these wavelengths by warm-blooded animals. The 3.5 to 14μ imagery embraced the appropriate wavelengths, but included such a broad spectrum of energy that we suspected a masking effect.

INFLUENCE OF VEGETATION

There are severe limitations on the applicability of the technique to animal census problems. One major difficulty is the fact that infrared radiation does not readily penetrate green foliage. This means that the method (at least with the present equipment) is useless for counting animals under a complete canopy of leaves. In the case of deciduous vegetation, census can be held after leaf fall. Evergreens, however, present an effective barrier.

In the George Reserve test, we placed individual deer in three small pens under different cover conditions; open grassland, oak *(Quercus velutina)* forest without leaves, and a closed canopy of red pine *(Pinus resinosa)*. The deer in the open pen and the one in the hardwood forest were clearly visible in the imagery. The deer under the pine canopy would not have been detected from the imagery if its exact location had not been known. The technique is thus most applicable to open range lands, tundra, low brushy areas, or defoliated deciduous forests.

VARIATIONS IN APPARENT TEMPERATURE

Another major difficulty is the variability in apparent temperatures of both the animal and the background depending upon surface characteristics, weather, and other factors. The apparent temperature of animals may be above, equal to, or below that of various objects in the environment. Furthermore, apparent temperatures of inanimate objects in the background will vary considerably. For example, a charred log will absorb and radiate heat in a far different manner than a granite boulder. These objects are of a size that could easily be mistaken for large animals on the imagery if they had similar apparent temperatures. Such features as roads, rivers, or lakes will also vary in apparent temperature, but these are readily recognized and should not confuse animal counts.

Ideal conditions for infrared censusing are those which result in uniform background temperatures with a maximum differential between animal and background. The success of the George Reserve trial was largely due to the nearly ideal conditions in which it was conducted. The experiment was held on January 4, 1967, at midday, with an air temperature of -4°C, and mild wind conditions. A 6 to 8 inch blanket of snow covered the area, and this, in conjunction with a high overcast sky, resulted in an extremely uniform background. Apparent temperatures, as measured by a hand-held radiometer, showed an approximately 7°C differential between the deer in small pens and the snow background. A scanner with a three milliradian IFV can detect about a 1°C temperature differential. Infrared scanners can detect very much finer temperature differentials (easily to 0.01°C) by increasing the size of the IFV, but at the expense of resolution. The three milliradian IFV and 1°C sensitivity are the result of a compromise between the need to resolve objects of large animal size and the need to maintain a reasonable temperature differential. Theoretically, very small animals could be detected by using a very small IFV provided there was a large temperature differential between the animals and their background.

The work of Marble (1967) has substantially increased our knowledge of the influence of several variables on the apparent temperatures of large animals and representative backgrounds. She took systematic radiometric readings on the ground of bison, *Bison bison;* antelope, *Antilocapra americana;* white-tailed deer. *Odocoileus virginianus;* and mule deer, *O. hemionus* and eight backgrounds: gravel road, uncut grass, lawn, juniper, *Juniperus scopulorum,* foliage and trunk, rock, log, and snow.

Marble's results agree closely with our actual aerial scanning. She found that under clear skies with a normal background that the differential in apparent temperature between animal and background was highly erratic. At times the apparent temperature of the animal was above, below, or the same as background objects. A snow background, by contrast, was extremely uniform in apparent temperature and gave the best contrast with the animals. However, even with a snow background, apparent temperature of an animal was influenced markedly by other environmental variables (Fig. 1). The question is whether this variation was due to changing air-temperature or to daytime versus nightime readings.

It appears that the variation in apparent surface temperatures of animals as measured by Marble (Fig. 1.), was primarily due to daytime versus nighttime influences. Temperature differentials were much greater during the daytime than at night. In Figure 2, the data for day and night overcast skies show two separate arrays. Similarly, her readings at night under a clear sky with no snow background showed the animal to be 5 to 6°C below apparent background temperature, while in the daytime they were interspersed.

The results of Moen (1968a) on surface temperatures of white-tailed deer in cold weather, during the day, under diffuse light are, perhaps, not directly comparable because of the use of different sensor equipment and methods. Yet, the regressions of animal temperature on air temperature derived independently by Marble and by Moen (1968a) compare reasonably well (Fig. 1). Moen's data showed a much closer fit, but this is probably due to the more detailed controlled conditions under which his readings were obtained. The conditions under which Marble took readings more nearly approximate the situation under which actual infrared scanning census is done.

Clearly temperature differential between animal and background is much greater with snow on the ground than under a cloud cover without snow. The differential in Marble's data under clouds without snow was about 3°C (Fig. 2), but there was considerable overlap, including times when the background was warmer than the animal. Without a snow background, the best conditions for infrared census is during a high overcast. It remains to be demonstrated that actual infrared animal counts can be successful under these conditions. Perhaps further work on the density of the overcast will clarify the influence of cloud cover upon apparent temperatures of animal and background. Presumably a cloud cover "shields" the earth from the radiation heat sink of the sky and reduces differential heating of objects in direct sunlight. Because it tends to stabilize apparent temperatures, a cloud cover may further enhance the differential between the animal and the ideal snow background (Fig. 2).

Although Marble had relatively few observations of the effect of wind on animal apparent temperatures, she suggested that it reduced

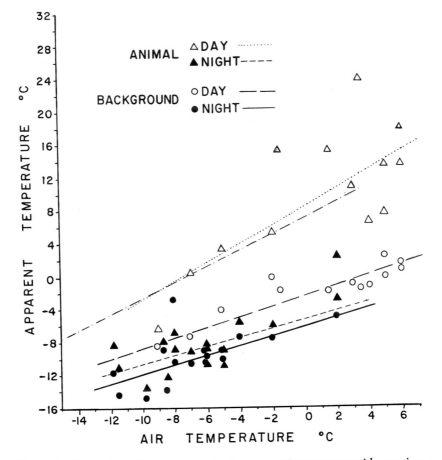

Figure 1. Comparison of minimum animal apparent temperature with maximum background apparent temperature composed of a blanket of snow. Data replotted from Marble (1967). Daytime (8:00 a.m. to 5:00 p.m.) and nighttime (6:00 p.m. to 7:00 a.m.) are based upon the average time of sunrise and sunset. The dot-dash regression line is from Moen (1968a) and was derived from daytime readings taken under diffuse light.

the differential between animal and background. This conclusion was substantiated by the work of Moen (1968b) who demonstrated that as convection heat loss increased, surface temperature and hence radiation loss, decreased. Thus, quiet conditions are most appropriate for infrared censusing.

Marble's observations support the well-known interference of atmospheric moisture in the form of mist, rain or snow with the transmission of infrared radiation. To summarize, ideal conditions for detecting large animals by infrared scanning are little or no wind, a snow background, high overcast sky, and a daytime overflight. If snow background is lacking, a day with high overcast and light winds may give successful

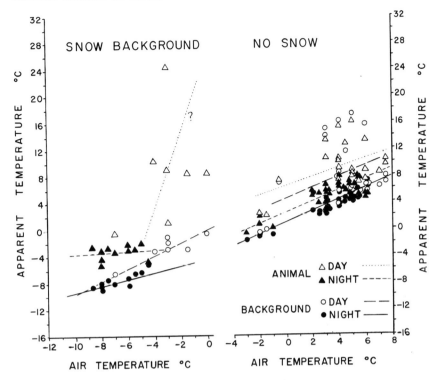

Figure 2. Comparison of minimum animal apparent temperature with maximum
    background apparent temperature with and without snow under a 100 percent
    overcast. Data replotted from Marble (1967). Daytime and nighttime as in
    Figure 1.

results, but the temperature differential between animal and background
would be quite narrow.

DISTINGUISHING SPECIES OF ANIMALS

While there are many situations where a single species of animal is
the only one likely to be recorded, mixed species are probably the more
common case. Furthermore, the question of how small an animal might
be detected complicates the issue. For example, it would be difficult to
get an accurate count of deer if rabbits, squirrels, and animals of
similarly small size produced comparable images.

One potential source of discrimination is the widely variable character
of the pelage between species which might give different apparent
temperatures. Croon (1967) took radiometric readings on white-tailed
deer, red fox *(Vulpes fulva)*, and red squirrel *(Tamiasciurus hudsonicus)*
and found little difference in apparent temperatures between them.
Similarly, Marble (1967) reported considerable overlap between the
apparent temperature of two species of deer, antelope, and bison. At

the time of the successful overflight of the George Reserve we obtained radiometric readings on wild turkeys and a man's wool coat and found them to be similar to readings from deer. Bare skin (a man's face) gave considerably higher apparent temperatures. The radiant heat loss of an object depends upon its emissivity and surface temperature (Kelvin scale to the fourth power) times the Stefan-Boltzmann constant. Hammel (1956) showed that most animals have an emissivity of .98 to 1.00, and therefore, function like black bodies. See Gates (this symposium) for a discussion of absorptivity of various animals. Because the characteristics of the coats of various wild vertebrates are similar, they tend to have comparable apparent temperatures under a given set of environmental conditions.

It seems unlikely therefore, that differences in apparent temperature will be useful in separating most species of birds or mammals in infrared imagery, with the exception of naked man. Because apparent temperatures of most animals are quite similar, Croon (1967) used the projectional area of the animal to determine its detection probability under the conditions of the successful George Reserve flight. He found that it would be extremely improbable that a fox- or raccoon-(*Procyon lotor*) sized animal would be detected with the equipment used. Obviously, detection of different sized animals would depend upon the size of the IFV, and the aircraft altitude.

A certain amount of tailoring of equipment and flight altitude to distinguish animal species may be possible with large animals; but the constraint of resolution versus sensitivity is such that the prospects are poor for small animals. Presumably a bison, which completely filled a three milli-radian IFV, could produce a hotter "spot" on the imagery than a deer, which would fill only about one-half of the field. However, differences in size of young and adult animals may prove to be an insurmountable problem. A bison calf may be closer in size to deer than to bison. Also, individual variation in apparent temperature between animals may be due to microhabitat and other factors which would severely complicate detection.

Another difficulty is that many small animals such as a group of roosting birds could produce an image similar to that of a single large animal. Such cases appear to be relatively uncommon in nature, for most species maintain an individual distance, or living space, great enough to provide separable images with a 3 milliradian IFV. Among large animals, the young may be kept virtually under the body of the mother, but suitable background conditions for scanning do not coincide with the season of birth.

It seems that there is little prospect with present equipment for separating species in the imagery. However, parameters can be set so as to "screen out" animals below a certain size. Problems of species composition of animals of similar size will have to be solved by spatial distribution of different species, or partitioning of a total infrared scanning count according to ratios determined by other methods.

ANIMAL BEHAVIOR IN RELATION TO INFRARED CENSUSING

Many of the large wild animals are nocturnal in habits. The use of thermal scanners is possible at night and would allow the overflight to be conducted at the time when many animals are most active. This has advantages particularly for species which tend to den or rest in heavy cover, particularly conifers, during the day. If it is desirable to census animals such as deer, the possibility of detecting intermediate sized animals (coyotes, foxes, badgers, raccoons, etc.) is greater at night than during the day when these smaller animals tend to den.

Animals bedded in exposed positions are recorded as well as actively feeding ones. On the George Reserve, deer in a dense, tangled tamarack swamp (probably bedded) were clearly visible on the imagery. Recently vacated beds may be warm enough to cause some difficulty by producing false signals in the imagery similar to those produced by animals. Marble (1967) reported that the apparent temperature of beds was 1 to 2°C warmer than the animal, but dropped below the animal temperature within one minute of the time the animal left it. Flying at 1000 feet altitude would cause little alarm to most animals; lower flight altitudes may record a number of recently vacated animal beds.

AVAILABILITY AND COST OF OPERATION

Most scanners have been designed and built under governmental contract, and are classified because of their military applications. Eventually these instruments will be declassified and available for use. Already the "M-2" scanner developed by the Willow Run Laboratories of the University of Michigan has been declassified and HRB-Singer has built one declassified scanner. However, neither of these scanners is readily available. The only scanner which is presently available on the market was designed and built by Bendix Corporation. Scanning equipment is expensive. The basic Bendix scanner costs in the neighborhood of $50,000. Of course, with large production, the cost per unit may decline.

Infrared scanners have a number of uses; it would probably be an unusual case where a scanner was used exclusively for animal census. A resource agency, for example, could use a scanner for vegetation and wetland mapping, monitering water pollution, and fire detection and control. Thus, the cost per application might be entirely reasonable.

An aircraft at least equivalent to a Cessna 180 is required to carry the present equipment. The cost of operation would include flight expenses and salaries of a pilot and an equipment engineer. Another cost which should be anticipated is the services of a skilled interpreter.

FUTURE PROSPECTS FOR INFRARED CENSUSING

From the above discussion it is obvious that the limitations of infrared scanning for census work are such that many large animal survey problems cannot be solved by this technique. In those situations where the peculiar requirements of the method are met, infrared will give a better count of animals over large areas than any other method presently available. There is every reason to believe that future generations of scanning hardware will be more sensitive and precise than those presently in operation. Further study of applied aspects, including animal census, is needed, but the time when infrared scanning is a routine option available to the resource manager is near.

SUMMARY

A successful count of the George Reserve white-tailed deer herd demonstrated the feasibility of censusing large animals with infrared scanners. Hardware filtered to 8 to 14μ is most appropriate for detecting warm-blooded animals. Scanners filtered to 3.5 to 5.5 and 3.5 to 14μ failed to detect deer. Application of the technique is possible in open habitat, or deciduous woody vegetation without leaves.

Detection of large animals depends upon a differential in apparent temperature between the animal and the background, both of which fluctuate due to surface characteristics and weather. Conditions giving the greatest temperature differential were a complete snow cover (which produces a uniform background), no wind, a high overcast sky, and a daytime overflight. As conditions depart from these, differences in apparent temperature decline, and indeed, the animal may appear to be colder than background objects.

Under a given set of conditions, most species of birds and mammals have similar apparent temperatures. Therefore, distinguishing between species with thermal scanners must be based upon size of the animal. Prospects for accomplishing this are poor, but the minimum size of animal detected can be controlled through careful specification of equipment parameters and flight altitude.

Other things being equal, daytime flights appear to be best for most census problems. Nighttime flights may be appropriate for nocturnal animals which go into coniferous cover or den during the day, but the differential in apparent temperature between animal and background is marginal at night, even under ideal conditions.

Most infrared scanning equipment is classified. Only the Bendix Corporation's scanner is commercially available. Probably, because of high cost, few agencies will obtain a scanner exclusively for animal census work; however, the equipment has wide applicability to other natural resource management problems.

REFERENCES

Croon, G. W. 1967. The application of infrared line scanners to big game inventories. Unpublished report, University of Michigan, School of Natural Resources. 37 typewritten pp.

Croon, G. W., D. R. McCullough, C. E. Olson, Jr., and L. M. Queal. 1968. Infrared scanning techniques for big game censusing. J. Wildlife Management 32(4):751-759.

Hammel, H. T. 1956. Infrared emissivities of some arctic fauna. J. Mammalogy 37:375-378.

Marble, Harriet P. 1967. Radiation from big game and background: a control study for infrared scanner census. Unpublished Master's Thesis, University of Montana, Missoula. 86 pp.

Moen, A. N. 1968a. Surface temperature and radiant heat loss from white-tailed deer. J. Wildlife Management 32:338-344.

Moen, A. N. 1968b. Energy balance of white-tailed deer in the winter. Trans. North American Wildlife and Natural Resources Conf. 33:224-236.

Suits, G. H. 1960. The nature of infrared radiation and ways to photograph it. Photogr. Engr. 26:763-773.

10

Portable Radios in Determination of Ecological Parameters of Large Vertebrates With Reference to Deer

R. Larry Marchinton

INTRODUCTION

A FUNDAMENTAL problem in the study of large wild vertebrates in their natural environment is locating the animals and obtaining information about them without significantly altering their behavior. For many years scientists have avoided the problem and serious study has often been limited to smaller animals which can be more easily studied under field or laboratory conditions. Vast amounts of good information have been obtained in this manner but it is becoming very apparent that it is necessary to conduct field studies to gain valid insight into the complexities of the natural ecology and behavior of many species. Many writers in recent years have emphasized the limitations or misinterpretations of data derived solely from laboratory experiments (Ardrey 1966).

The development of transistorized radios for use on animals opened new opportunities for field studies beginning in the late 1950's. Progress, however, has not proceeded at the rate which was expected or in proportion to the number of researchers involved. The preponderance of technical papers on telemetry in ecological studies have concerned themselves with the technique and have presented relatively little new ecological information. This can be attributed to several factors. First, the development of good telemetry equipment has been slow due to insufficient funds for basic research and development. A communication barrier between scientists and technicians in the widely separated discipline of electronics and ecology is a related problem. It is very unusual for an individual to have sufficient expertise in both disciplines to develop equipment for his specialized needs and to apply

148

this equipment to ecological problems efficiently. This problem is beginning to be resolved as more electronic technicians become available with experience in meeting the special design and construction problems required of biotelemetry systems, and as ecologists become more cognizant of electronic capabilities.

A second basic problem is that researchers tend to expect the equipment to take the physical work out of field investigations. This may be true to some extent in the case of automatic systems such as those described by Cochran, et al. (1965). Although these are powerful tools for studying animal ecology and behavior, automated systems cannot be efficiently used to answer some ecological questions because of their relative immobility and high cost. Portable equipment such as that discussed in this chapter is available and can be obtained at reasonable cost. To maintain maximum utilization of this type of instrumentation, however, requires considerable physical effort and perseverance. It necessitates working throughout the daylight as well as the night hours, at least when studying species with variable activity cycles. Relatively few researchers seem willing to invest the effort required to realize the potential that portable systems now provide.

A final problem is that researchers expect too much from the equipment in terms of dependability and performance, and consequently quickly become discouraged. I have found it useful in organizing a project to define the minimum instrument performance criteria which can be used and still obtain useful information. When this is established as an original goal it is easier to realistically evaluate the success or failure of the equipment.

This paper will discuss applications and results of portable radio equipment in a study of a large terrestrial vertebrate. Although the techniques discussed were worked out for a deer study in the southeastern United States, they should be applicable to other species and in other regions.

INSTRUMENTATION

Telemetric instrumentation involved a wide variety of radio equipment including small transmitters that were placed on the deer and specially designed receivers which the investigator used to obtain information about the transmitter-tagged animals. Other equipment used included various directional and non-directional transmitting and receiving antennae and weather recording instruments. Much of the equipment is now commercially available and has been described in the literature (Cochran 1967, Cochran and Hagen 1963, Cochran and Lord 1963, and Tester, et al. 1964).

Transmitters

A total of 49 transmitters which were variable in performance, design, and weight were placed on deer. Two different frequency bands

were employed, (26 to 27mc and 150 to 153mc). Transmitter life
varied from a few hours to nearly seven months. Maximum monitoring
distances of the transmitter-receiver systems ranged up to twelve miles
but averaged about one mile depending on the system, terrain, and
weather conditions. Weight was a problem in some of the earlier
designs as the battery-transmitter combination weighed as much as 1.8
pounds. This weight necessitated the use of a harness to hold the
radios on the animal, since they were too large to conveniently carry
on a collar. After the early developmental stages the transmitter and
power supply were reduced to a few ounces and could be easily mounted
in a collar (Fig. 1). These collar-mounts proved very successful and
most of the animals were instrumented in this manner. Transmitters
operating on 27mc had a loop antenna built into the collar and since
the loop had to remain at a fixed circumference the collar was not
adjustable. The 150mc transmitter had a whip antenna and the collar
itself could be adjusted quite readily.

The transmitters were coated with colorful paint in various patterns
for the purpose of visual identification of the radio-equipped animals.
It was found valuable to use light-reflecting paint to facilitate night
identification. A tag offering a reward to any person returning a radio
was also affixed to the collars.

Effect on behavior

There was a tendency for deer movements to be restricted for a

Figure 1. A young buck wearing a collar-mounted 150mc radio transmitter. This
method of attaching the radio to the animal was very successful and most
radios were mounted in this way.

period of time immediately after the radio was attached. This period was usually less than one week but varied depending on the way the animal was captured. Deer that were shot with a drug gun usually had a longer period of adjustment than those captured in a box trap. In general, observations of deer wearing the collar-mounted transmitters indicated that they were not disturbing the animals in any observable way. This was not the case with harness mounted radios.

Receivers

Both portable and semi-portable receiving equipment were used. Portable receivers were light enough to be carried into the field with their power source (Fig. 2). The semi-portable receivers were operated from the vehicle battery and therefore had to remain near the vehicle. Receivers dependent on vehicle power were supplemented with a small portable receiver to approach the animal for visual observation.

Several different directional and non-directional antennae were used in conjunction with the receivers. Non-directional antennae included whips with lengths corresponding to $\frac{1}{4}$ of the transmitted wave length. These antennae were mounted on a vehicle and used initially to locate the animal's radio signal. The direction of the signal could then be ascertained with a directional antennae. The loop and yagi

Figure 2. One of the portable receivers used in the study. (Photograph by Gulf States Paper Company personnel.)

were the two basic types of directional antennae employed in this study. The former directionalized best from the "null" or no-signal position, while the latter used the maximum signal and a visual display indicated the direction. Loop antennae were used with 26-27mc and yagi antennae with 150-153mc systems. Semi-permanent yagi antennae were set up in some locations to increase the range of signal reception and to facilitate detection of slight movements.

TELEMETRIC PROCEDURES

Remote tracking in natural environment

Seven white-tailed deer *(Odocoileus virginianus)* populations distributed over four southern states were selected for study. These locations were selected to represent a wide variety of ecological conditions. Deer were captured with box traps or the Cap-Chur Gun technique (Thomas and Marburger 1964) and the animals were weighed, measured, aged, and sexed before being radio-instrumented. After installation of the transmitter on the deer, the transmitter-receiver system was thoroughly tested. The animal was then released at the capture location unless it was to be introduced into a new habitat.

The radio tracking procedure was variously modified to meet local conditions on each study area. Generally the deer's range was canvassed using a vehicle equipped with a whip and/or directional antenna and a mobile receiving unit until a strong radio signal was received. A directional reading was then taken and recorded by compass azimuth at the nearest predetermined station. A null signal in the case of the loop antennae used in the 27mc systems or a maximum signal for the yagi antennae in the 150mc systems indicated the direction of the radiating signals as perpendicular to the elements of the antenna. The operation was repeated at a second station, with a minimum of time between readings, and the actual location plotted on a reference map by triangulation. These stations were given permanent identification numbers in the field and on a map, and new ones were added when the animal moved out of range of those already established. The number of stations necessary to track individual deer ranged from as few as three in semi-mountainous areas to as many as fourteen in level, forested terrain.

Readings were made at intervals of from 1 to 3 hours throughout a 24-hour tracking period. In most cases, individual deer were monitored for at least two complete 24-hour tracking periods each week. In other instances animals were monitored every two hours for continuous periods greater than a week. The 24-hour or diel period was considered the basic monitoring unit and the data were organized according to diel periods. This periodic monitoring usually continued until the transmitter failed, i.e. up to six months.

Sometimes three readings or more were obtained as a means of checking accuracy. On other occasions after the readings were taken an animal was tracked to its location and observed. Through these techniques it was ascertained that radio-locations averaged less than 75 yards from the animal's actual position. Greater accuracy could be obtained if the animal was less than 0.25 mile from the monitoring stations.

The elapsed time between radio fixes when the investigator moved from one station to another was a potential source of positional error when the animal was moving rapidly. This could be eliminated by using two receivers and a communication system so that directions could be obtained simultaneously. In actual practice, this was not considered necessary since the radio signal characteristics indicated whether the animal was bedded, feeding slowly or moving rapidly through the woods.

While, in general, the technique functioned adequately, a few technical problems arose. False directions were sometimes obtained from nearby telephone or electrical wires which could result in tracking errors. This was easily eliminated by moving away from the wires. Difficulty was also encountered in obtaining a direction when the transmitter was very close (under 75 yards) to the receiver. This problem was eliminated when the investigator developed experience in use of the receivers. Considerable signal "bending" or "bounce" was attributed to vegetation or terrain with the 150-153mc equipment. This could be minimized by restricting readings to hills and ridges.

Recording activity

In addition to spatially locating the deer at intervals in time, most telemetry systems used in this study simultaneously telemetered information concerning the animal's activities. With experience the investigator was able to detect characteristic patterns of variation in the telemetered signal and to associate these with the different levels of the animal's activity. This ability was most easily obtained by prolonged visual observations of the radio-equipped animal while receiving the telemetered signal to determine which movements were associated with particular signals. Based on this experience the animal was recorded as active or inactive at each location with other behavioral information being recorded when it was considered important.

Visual contact was made with the radio-equipped animals when it could be done without risk of disturbing them. For example, if the telemetered information indicated the deer was moving into an open area or across a road, the investigator would position himself to observe the animal. In this way direct observational data were obtained concerning social behavior, grouping, and related activities.

Recording meteorological information

Certain meteorological variables were thought to have important relationships to movement and activity. Measurements or estimates of wind velocity, cloud cover, temperature, humidity, precipitation, and barometric pressure were recorded on the data sheets in conjunction with each radio location and activity reading whenever possible. In some cases meteorological data were measured and recorded continuously. A hygrothermograph was set up in some instances and information was obtained from continuously recording United States Air Force research weather instruments located on one study area. Where precise instruments were not locally available, estimates were made, or information was obtained from the nearest ESSA weather station.

QUANTIFICATION OF DATA PARAMETERS

The first step in analyzing the data was the compilation of all locations, movement patterns, and related information obtained for each radio-instrumented individual. An analysis of this information was made to determine if the animal had been sufficiently monitored so that his home range and movement patterns could be described. This determination was primarily based on the length of time and frequency of the periodic radio contact after the animal's adjustment. In general, if an animal's home range was not increasing in size with additional locations it was assumed that the minimum home range was probably circumscribed, at least for the season of observation. In most cases this required at least one month of radio contact.

The data obtained on less than half of 49 instrumented deer were considered adequate to define normal movement patterns. The remainder of the animals, however, contributed to a better understanding of movements and activities even though their patterns were not completely defined. Analysis of the 17 most intensively studied individuals involved the establishment and quantification of certain movement and home range parameters. The following were included.

1. *Minimum home range.*—The area included within a line connecting the outermost radio locations of the deer during the entire period of telemetric and visual contact was referred to as the minimum home range. Since some of the ranges were irregularly shaped, an attempt was made to connect locations with lines that would result in the most nearly accurate home range acreage (Fig. 3). The technique is similar to the modified minimum area method described by Harvey and Barbour (1965), but differs primarily in that a knowledge of the habitat and the animal's movement patterns rather than a mechanical procedure was used in determining which points to connect in establishing the minimum home range boundaries. Although this method was somewhat subjective, it capitalized on the investigator's knowledge of the vegetation, physical barriers, food sources, and the animal's habits, and was probably more easily accurate than if a strict mechanical approach had been used.

2. *Home range major axis.*—A line segment formed by connecting the two radio locations of the deer, obtained any time during the

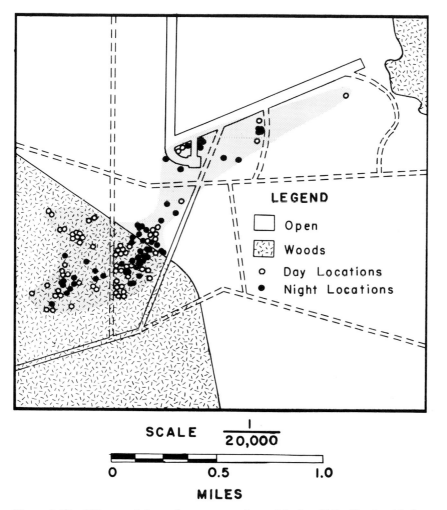

LEGEND

☐ Open

▨ Woods

○ Day Locations

● Night Locations

SCALE $\dfrac{1}{20,000}$

0   0.5   1.0

MILES

Figure 3. The 200-acre minimum home range of an adult doe, Eglin No. 2, with day and night locations indicated. She was studied telemetrically from June 24 to August 17, 1964, when she died while giving birth.

study, that were the greatest distance apart constituted the home range major axis. In instances where such a line would not lie entirely within the minimum home range, it was angled such that it would approximate the mid-line of the range and still connect these two radio locations.

3. *Home range minor axis.*—A line segment perpendicular to the major axis and connecting the boundaries of the minimum home range at its widest point was the home range minor axis.

4. *Distance between extreme diel locations (DBE)*—The greatest distance between any two radio locations of the deer during a par-

ticular 24-hour tracking period was its DBE for that period (Fig. 4).
The average and range of DBE values were determined for individuals,
population samples and total sample, and were studied in relation to
season, habitat, meteorological, and other variables.

5. *Minimum total distance moved in diel period (MTD)*—The sum
of the distances between sequential locations of an individual deer
during a particular 24-hour period of tracking was its MTD for that
period (Fig. 4). The data upon which this value was based were subject
to a certain amount of experimental error because of variation in number
of radio locations and time intervals between stations. The average

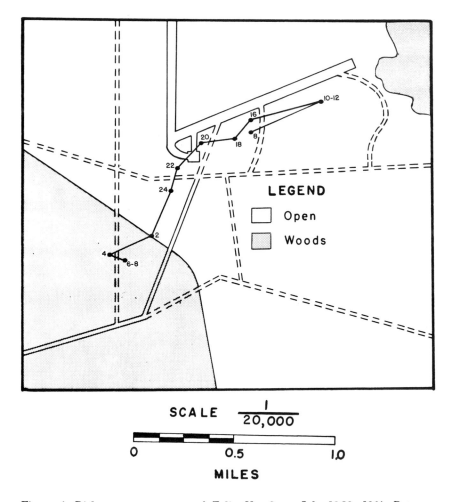

Figure 4. Diel movement pattern of Eglin No. 2 on July 18-19, 1964. Dots are
    radio-locations and numerals are the hours (based on the 24-hour clock) during
    which the animal was located. The DBE and MTD for this period were 1.40
    and 2.23 miles.

and range of MTD values were compiled and studied as described for the DBE parameter.

6. *Core area.*—According to Kaufmann (1962), the "core area" is that area within the home range of an animal or social group which is used most frequently and this area probably contains the principal home sites, refuges, and most dependable food sources. Based on this definition, most deer had one or more definable core areas within their home range. These were indicated on home range maps by clusters of radio locations and were discussed in terms of percentages of the total home range area and time spent in them as well as their relative stability.

7. *Home range length-width quotient.*—The quotient of the minimum home range major axis divided by the minimum home range average width is the home range length-width quotient. Average width is defined as the mathematically determined value found by dividing the minimum home range area expressed in area units by the minimum home range major axis expressed in linear units. This parameter is an expression of the length to width ratio.

For example, the movement parameters and "case history" of a typical individual are listed below. A doe 2.5 years old, designated Eglin No. 2, was captured and radio instrumented on June 24, 1964. This animal was in good physical condition, weighed eighty-five pounds, and appeared to be pregnant. Telemetric monitoring of movements continued periodically until the animal was found dead near the center of its home range core area on August 17, 1964. A post-mortem examination indicated she had died during parturition.

Her minimum home range was 200 acres and is shown in Figure 3. It had a major axis of 1.80 miles and the minor axis only 0.40 miles. The range was greatly elongated as indicated by its length-average width quotient of 10.37 and was oriented in a northeast-southwest direction. The individual DBE values were quite variable, ranging from 0.30 miles on June 29-30 to 1.40 miles on July 18-19, (see Figure 4), but had an arithmetic mean of 0.91 miles. Excluding the restricted movements during the period immediately after installation of the radio, the minimum total movement during each diel period (MTD) ranged from 0.90 miles to 2.23 miles and averaged 1.70 miles.

Although the behavior of Eglin No. 2 varied considerably, a common pattern involved bedding and feeding in the wooded portion of the habitat during the daylight hours and moving onto the open habitat during the late afternoon or night. Often the middle and latter parts of the night were spent bedding out on the unwooded habitat. Some diurnal bedding occurred on the open range also, but this was usually in a section where turkey oak reproduction offered shelter. Most of the diurnal activity was in the home range core area, which was primarily wooded.

DIRECT OBSERVATION OF INSTRUMENTED,
BEHAVIORALLY CONDITIONED DEER

Many kinds of ethological and ecological information cannot be obtained entirely by remote sensing with the present level of instrument sophistication but require direct observation of the subject. The most commonly used method of studying the food habits of mammals, for example, has been the examination of food remains in the digestive tract or feces. Such studies were adequate to estimate gross food intake but were difficult to correlate with food availability because the investigator usually lacked detailed knowledge of foods available at the time and place the animal fed. Ideally, knowledge of food preferences and many other aspects of behavior can be best obtained by direct observation of the animal at close range.

An early report (Cory 1927) of detailed cattle feeding behavior was based on direct observation. He recorded cattle activity including feeding by minutes, and reported the results as minutes of activity per 24-hour period. This method has been applied to wild ungulates by Dixon (1934), Hahn (1945), White (1961), and Buechner (1950). Wariness of wild animals, however, makes data collection by this method very difficult. McMahan (1964), Wallmo (1964), and Watts (1964), each reported the use of fully tamed deer for direct observation of their feeding habits. These studies illustrate the potential of this technique for detailed availability-utilization studies, but leave open to question the effects of close supervision and control on the behavior of the subject.

Specially conditioned deer and portable telemetry equipment were used in the present study to obtain information concerning aspects of behavior, including food habits, that could not be readily obtained using the techniques previously described. An earlier paper (Marchinton and Baker 1967) discussed this telemetric-observational technique; it involves the use of a combination of radio-location and direct observation of deer which have been conditioned to tolerate the close proximity of an observer, but have subsequently been allowed to establish themselves as functioning units in a natural ecosystem.

It was considered important for the animals employed, to be conditioned for a reduced "flight distance" of from 5 to 15 yards. An equally important criteria was that the animal not become so "tame" that it would be attracted to the observer.

Although different procedures could be used to condition animals for this type of work, the following sequence was used in the present study. The animals were: 1) born in captivity or captured at less than one week of age; 2) separated from the doe, hand fed, and given close human contact during the first month of life; 3) relatively but not completely isolated from humans after this period; 4) held in moderate-sized enclosures (1-3 acres) where some selection of natural foods was possible until ready for release; 5) equipped with a radio transmitter, preferably at the age of 10-12 months and released into a habitat supporting a wild deer population; and 6) allowed a few weeks to become adapted to the new environment before study proceeded.

The usual procedure for obtaining data was to locate a deer by radio telemetry and then to walk in the direction indicated until visual contact was made with the animal. The investigator remained near enough to identify the material consumed and observe other pertinent behavior for several hours at a time, periodically during the week. The length of time in minutes spent feeding on each food item was recorded in a manner similar to the feeding-minutes method described by White (1961). Other information including defecation rate, rumination cycles, bedding and movement patterns, activity cycles, social behavior, and behavioral interaction with other species, particularly free-roaming or wild dogs, were also recorded.

The primary purpose of this chapter is to describe techniques; therefore, a detailed account of the results is not given. The following information concerning a one-year-old-buck studied during the summer of 1964 is presented to illustrate the usefulness of the technique. Although radio contact was maintained periodically for nearly four months, a total of only 49 hours on sixteen days between June 22 and July 30 was spent recording food habits data. This was felt to be an adequate sample of the animal's food habits during the period. During this time, 44 different types of plant material were observed being consumed. The ten most important items according to percentage of time spent feeding on each were: many-flowered aster (*Aster* sp.), 37.9; alfalfa (*Medicago sativa*), 20.5; wild plum fruit (*Prunus* sp.), 7.9; blackberry foliage (*Rubus* sp.), 6.6; various grasses, 6.1; blackberry fruit (*Rubus* sp.), 4.3; honeysuckle (*Lonicera japonica*), 2.6; sorrel (*Rumex* sp.), 1.8; wood sorrel (*Oxalis* sp.), 1.7; beggar's tick (*Desmodium* spp.), 1.4. Defecations usually occurred while the animal was actively feeding and averaged about one defecation for each two or three hours of visual contact but were not evenly distributed through time. Often one defecation was followed by another about 20 minutes later. Fecal material tended to be loose and individual pellets were seldom formed. Further studies with other animals indicated that was the result of the quantities of fruits and succulent foliage being consumed during the spring and summer months.

Feeding periods were followed by periods of bedding and rumination lasting from a few minutes to several hours. Bedding locations were scattered over a 255-acre home range but a few areas seemed to be preferred and he often traveled several hundred yards to reach these places.

The telemetric-observation technique has many advantages over observation of fully wild deer in their natural environment because of the difficulty of approaching the latter close enough to observe habits in detail. It is, of course, recognized that in the process of conditioning the animal for the desired traits, we also may be conditioning him to respond in an unusual manner to other factors in his environment. Continued study should indicate to what extent this is true.

ECOLOGICAL PATTERNS AND THE VALUE OF TELEMETRY

A fundamental concept of ecology and behavior is that for most species of animals there tends to be a characteristic movement pattern and activity cycle. In general, these behavioral characteristics can be explained in terms of their selective advantage for the species' survival in a complex ecosystem. For example, an animal must be active during the time when he can best obtain the necessities for life and reproduction and when he is in the least danger from deleterious factors in the environment. Predation and the availability of food are probably the chief ecological factors controlling the mammalian 24-hour activity cycle except in desert regions, according to Cloudsley-Thompson (1961).

It is to the animal's advantage in the process of natural selection, to establish a movement pattern which is repetitious enough to provide familiarity with a particular unit of habitat. The animal is then able to obtain its basic necessities of life, such as food, water, and protection with maximum efficiency and minimum energy expenditure. Probably the size of an animal's home range is governed by two requirements:
1. It must be large enough to provide all of the basic essentials for life and reproduction.
2. It must be small enough so that the animal gains some advantage from his familiarity with the area.

The nature of cyclic activity in deer is of interest from a practical, as well as an academic, point of view. To the field biologist and wildlife manager it is of value to understand the activity cycle of economically important species and how they change with season of the year, weather, and disturbances from man and other predators. By understanding the cycle one can anticipate activity, thus facilitating field work with these animals.

Knowledge of home range and daily movement is probably even more important for practical wildlife management. One of the most commonly used deer population census methods in the southeastern United States is the track-count technique developed by Tyson (1952). The reliability of this method is related to the accuracy of the estimate of daily movement of deer in the area in which the census is made.

Although the need for this information in practical game management is apparent, it may be of even greater importance for more fundamental scientific reasons. According to the anthropologist, Robert Ardrey (1966), the tendency for animals to demonstrate behaviorally their attachment to a unit of habitat has important significance in terms of human behavior. He states that this "attachment" may have been the most important single factor in the evolution of social behavior in man as well as lower animals. Recently many psychologists, sociologists, and anthropologists have become interested in this theory and studies of animal behavior which relate to it. In view of the important implications involved it would behoove zoologists to place emphasis on obtaining very detailed information concerning the ways animals disperse and relate themselves to their environment within the dimensions of space and time.

Radio telemetry seems to provide many opportunities for study of the intricate ecological, evolutionary, and genetic questions involved in the formation of movement and activity patterns. By quantifying the movement parameters of individual deer ecological patterns became evident which could be related to their sex, age, reproductive condition and social interactions, as well as, to the habitat, and population of which they were a part, and to meteorological variables in their environment. Furthermore, study of home range parameters through a few generations of individuals has suggested evidence as to the actual ontogeny of the home range phenomena (Jeter and Marchinton 1964, Marchinton and Jeter 1966, and Marchinton and Baker 1967).

Radio telemetry is a comparatively recent innovation in the field of ecology. Many problems have been encountered with techniques and the equipment eventually used was not ideal. The need for improvement is evident in such areas as transmitter life and dependability as well as receiver sensitivity. Nevertheless, the telemetric systems were considered adequate for further use in studies of this type. The results have indicated that commercially available equipment can yield unique and valuable information. The worth of such data can hardly be overestimated when the man-hours required by other methods to gather similar information are considered. Without telemetry much of this information could not have been obtained, and the researcher should not overlook opportunities to utilize advances in electronic technology. If ecology is to continue progressing rapidly as a science, the application of such interdisciplinary techniques is essential.

SUMMARY

Portable radio tracking equipment in the study of the ecology of large vertebrates with reference to white-tail deer in the Southeastern United States was discussed. During a six-year period 49 radio transmitters were placed on deer in seven different habitats including river bottom, sandhill, piedmont and semi-mountainous terrain. Animals were equipped with collar-mounted transmitters and periodically located by triangulation with portable receivers. Data were recorded in 24-hour increments and analyzed for diel movement and activity patterns. A number of different parameters were developed to give quantitative and descriptive values to movement and activity patterns. Quantitative comparisons between individuals and population movement parameters indicated relatively uniform home range size inspite of habitat differences, although there were indications that home range size may be an inverse function of population density. Using these data it was possible to relate variation in parameters to the animal's age, sex, and reproductive condition as well as to the habitat, population, and meteorological conditions in the environment. A technique involving observation of radio-tagged, specially preconditioned animals was also discussed and some sample results given.

REFERENCES

Ardrey, R. 1966. *The Territorial Imperative.* Antheneum, New York. 390 pp.

Buechner, H. K. 1950. Life history, ecology, and range use of the pronghorn antelope in Trans- Pecos Texas. Am. Midland Naturalist 43:257-354.

Cloudsley-Thompson, J. L. 1961. *Rhythmic Activity in Animal Physiology and Behavior.* Academic Press, Inc., New York. 236 pp.

Cochran, W. W. 1967. Some notes on the design of a directional loop antenna for radio-tracking wildlife. AIBS/BIAC Information Module M4.

Cochran, W. W. and T. E. Hagen. 1963. Construction of collar transmitters for deer. Univ. of Minnesota, Museum Nat. Hist. Tech. Rept. 4:12.

Cochran, W. W. and R. D. Lord, Jr. 1963. A radio-tracking system for wild animals. J. Wildlife Management. 27:9-24.

Cochran, W. W., D. W. Warner, J. R. Tester, and V. B. Kuechle. 1965. Automatic radio-tracking system for monitoring animal movements. BioScience 15:98-100.

Cory, V. L. 1927. Activities of livestock on the range. Tex. Agri. Expt. Sta. Bull. No. 367:47.

Dixon, J. S. 1934. A study of the life history and food habits of mule deer in California. California Fish and Game 29:315-154.

Hahn, H. C. 1945. The white-tailed deer in the Edwards Plateau region of Texas. Rept. Fed. Aid. Proj. I-R, Tex. Game, Fish and Oyster Comm. 52 pp.

Harvey, M. J. and R. W. Barbour. 1965. Home range of *Microtus ochrogaster* as determined by a modified minimum area method. J. Mammalogy 43:398-402.

Jeter, L. K. and R. L. Marchinton. 1964. Preliminary report of telemetric study of deer movements and behavior on the Eglin Field Reservation in northwestern Florida. Proc. 18th Ann. Conf. S. E. Assoc. Fish and Game Comm. (In Press).

Kaufmann, J. H. 1962. Ecology and social behavior of the coati, *Nasau narica,* on Barro Colorado Island, Panama. Univ. California Publ. in Zool. 60:95-222.

Marchinton, R. L. and L. K. Jeter. 1966. Telemetric study of deer movement ecology in the Southeast. Proc. 20th Ann. Conf. S. E. Assoc. Fish and Game Comm. pp. 189-206.

Marchinton, R. L. and M. F. Baker. 1967. Food habits study of deer by means of a telemetric-observational technique. Proc. of 44th Ann. Ala. Acad. Sci.

McMahan, C. A. 1964. Comparative food habits of deer and three classes of livestock. J. Wildlife Management. 28:798-808.

Tester, J. R., D. W. Warner, and W. W. Cochran. 1964. A radio-tracking system for studying movements of deer. J. Wildlife Management. 28:42-45.

Thomas, J. W. and R. G. Marburger. 1964. Mortality in deer shot in the thoracic area with the Cap-Chur Gun. J. Wildlife Management. 28:173-175.

Tyson, E. L. 1952. Estimating deer populations from tracks: A preliminary report. Pres. 6th Ann. Conf. S. E. Assoc. of Game and Fish Comm., Savannah, Georgia 15 pp.

Wallmo, C. 1964. Arizona's "educated" deer. Ariz. Game and Fish Department Wildlife Views 11:4-9.

Watts, C. 1964. Forage preferences of captive deer while free ranging in a mixed oak forest. M.S. Thesis. Penn. State Univ. 65 pp.

White, R. W. 1961. Some foods of the white-tailed deer in southern Arizona. J. Wildlife Management. 25:404-409.

11

The Significance of Spectral Changes in Light Scattered by the Sea

GEORGE L. CLARKE

INTRODUCTION*

STUDIES of the penetration of light into the ocean, particularly in relation to the photic reactions of plants and animals in the sea, have been carried out over a period of years (Clarke and Denton 1962; Clarke and Kelly 1965; Clarke 1965 and 1967). Radiant energy supplied from the sun makes possible the photosynthesis of the green plants which, for the seas as a whole, consist primarily of the phytoplankton. The production of all other types of marine organisms at successive trophic levels depends directly or indirectly upon the growth of green plants. A knowledge of the strength and nature of the light factor has therefore been recognized as crucial for an understanding of the operation of the ocean as an ecosystem.

As light penetrates into the sea, it is modified not only by the water through which it passes, but also by absorption and scattering due to the particulate and dissolved materials present. Of these, one of special prominence and importance is the phytoplankton itself and especially the chlorophyll and other pigments which it contains. Recent investigations have suggested that the modification of the light by the materials in the water may serve as a means for identifying different water masses. Areas found to contain high concentrations of chlorophyll would be known to contain large populations of phytoplankton and hence to be regions of high potential productivity. In the present report, the possibility of using this new procedure for the rapid delineation of water masses from the air is examined.

LIGHT INTENSITY

Our early measurements made with a relatively insensitive photovoltaic photometer showed that the depth to which daylight could

*Contribution No. 2151 from the Woods Hole Oceanographic Institution. Research supported by NSF grant #2435.

penetrate before its intensity was reduced to 1% of its surface value ranged from about 100 meters in the clearest ocean water to 20 meters or less in clear coastal waters and to still shallower levels in the more turbid waters of bays, harbors, and estuaries, and over fishing banks. One percent of incident daylight is approximately the threshold for effective photosynthesis. The photic responses of animals can be controlled by light intensities of far smaller magnitudes and thus can take place at greater depths and under twilight or nighttime conditions. To investigate these responses, it was necessary to develop deep-sea photometers containing photomultiplier tubes. With such instruments, the conditions of light in the water have been investigated in a great variety of regions both close to shore and in the open ocean. In Figure 1 a curve is shown indicating the penetration of daylight (sun plus skylight) into the clearest ocean water in relation to thresholds of biological importance, and a second curve gives the same information for clear coastal water. The intensity of full moonlight at the surface is seen to be about 6 orders of magnitude below that of daylight, or about 10^{-1} $\mu w/cm^2$. The intensity of the night sky when clear is about 10^{-4} $\mu w/cm^2$, and when heavily clouded is about $10^{-5} \mu w/cm^2$. Also indicated on the diagram is the strength of upward-scattered or "upwelling" light. This is seen to have an intensity of about 1% or 2% of the "downwelling" light. Some of the scattered light passes upward through the surface of the water and can be recorded by a photometer suspended over the water from the boom of a ship, from an airplane, or from a satellite. The light scattered upward from the subsurface layers must be distinguished from any radiation reflected from the water surface itself. The general level of intensity of upwelling visible radiation from beneath the surface will differ according to the clarity of the water, and the spectral distribution of the upwelling light will be further modified by the nature and amount of particulate and dissolved material in the water.

LIGHT QUALITY

As total daylight (sun plus skylight) reaches the sea surface through air mass 1, its highest intensity occurs at a wavelength of about 490 mµ. As the sun approaches the horizon, the red component of its direct radiation becomes relatively stronger, but since a larger share of the incident light now comes from the sky, the blue portion of the spectrum still remains the strongest for the total radiation.

As the incident radiation penetrates into natural waters, the infrared component is removed within a few meters. The far ultra-violet is also attenuated rapidly, but the near ultra-violet penetrates more effectively. The highest rate of penetration occurs in the visible part of the spectrum. Within this region, red and yellow are attenuated rapidly, green is attenuated more slowly, and the blue component is the most penetrating in the clearest waters.

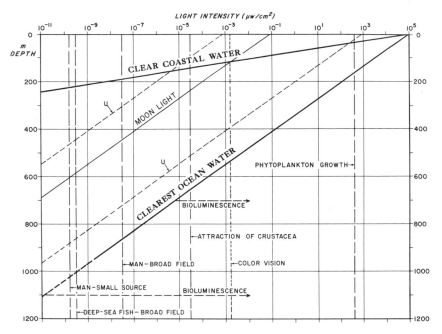

Figure 1. Schematic diagram to show the penetration of sunlight into the clearest
    ocean water (k = 0.033) and into clear coastal water (k = 0.15) in relation
    to minimum intensity values for the vision of man and certain deep-sea fishes. The
    approximate minimum values for the attraction of crustacea, colour vision in man,
    and for phytoplankton growth are indicated as well as the range of intensity of
    bioluminescence in the sea. The penetration of moonlight and the approximate
    intensities of upward scattered (u) sunlight and moonlight in the clearest
    ocean water are also shown. Since light penetrating deep into the sea is
    confined to a narrow band of wavelengths, the values given for colour vision
    represent the approximate maximum depths at which a blue hue would be
    observed. (Modified from Clarke & Denton, 1962).

Using a double monochromater, Smith and Tyler (1967) measured
the attenuation coefficient for radiation between 350 mμ and 700 mμ
in the extremely clear waters of Crater Lake, Oregon. Their curves show
the rapid reduction in the intensity of light at the ends of this spectral
range with the minimum attenuation occurring at 420 mμ.

Information on the added effects of all kinds of materials on the
spectrum of light in the water is desirable, but of prime importance
is the influence of chlorophyll since this is an indicator of the amount
of living plant material present. Calculations of the effect of increasing
amounts of chlorophyll in a natural water body first shows an increased attenuation in the blue region where the water itself is clearest.
With higher concentration of chlorophyll, the characteristic sharp
peak at 675 mμ begins to appear, and at the same time the absorption
in the blue region becomes greater and wider in its effect. As a
result, with increasing concentrations of this pigment the region of
maximum transparency moves from about 470 mμ to 550 mμ and
beyond.

When the upwelling light was measured by Smith and Tyler (1967) at depths of 5, 15 and 25 m in Crater Lake, the spectrum was seen to have been modified by the longer path length through the water with the result that the intensity of light at the longer wavelengths was further reduced. However, the maximum intensity remained at about 420 mμ and there were no regions where extra attenuation would indicate the presence of foreign materials. But, when similar observations were made in coastal marine areas, Tyler and Smith (1967) found marked modifications of the spectrum. At a location in the Gulf of California, the point of maximum intensity of downwelling light at 9 m and 14 m had moved to a wavelength of about 565 mμ and a region of extra attenuation occurred at about 670 mμ. Since the latter is near an absorption peak of chlorophyll, it is reasonable to assume that the dip in the curve was caused by a concentration of phytoplankton in this body of water. The upwelling light at these same depths exhibited similar peaks and troughs. Even at a depth of only 4 m the upwelling light which had been scattered to this level from deeper layers showed the same characteristic modification of the spectrum. In another series of measurements by these same investigators off San Diego, the spectrum of the upwelling light was found to reach a maximum at about 485 mμ and a minimum at about 650 mμ.

Preliminary investigations of the spectral changes in upwelling and downwelling light at localities along the east coast of the United States were carried out during the summer of 1967 by the author in collaboration with Dr. Gifford C. Ewing and Mr. Alfred C. Conrod, assisted by Garry F. Mayer, Robert M. Alexander, Foster L. Striffler and Eric Mattson. Two radiometers loaned by the Experimental Astronomy Laboratory of MIT and described by Conrod and Kezar (1966) were employed. One instrument was placed in a watertight and pressure-resistant case for use from our research vessels and the other was mounted in our research airplane. The energy was recorded on a Sanborn strip chart recorder for each of 25 narrow bands (10 to 15mμ wide) between wavelengths of 360 mμ and 650 mμ. Repeated scans of the spectrum were recorded, each complete scan requiring less than one second. The spectrum of the incident sun plus sky radiation was obtained by recording the light reflected from an 18% neutral grey card held in the open above the water surface.

The first set of observations reported here was made from the ship only and was carried out at a point in Woods Hole Harbor where the water is 25 m deep. The records of downwelling light (Fig. 2) show that already at 4 m the point of maximum radiation has moved to 545mμ. At 6 m, peak intensity is found at 560 mμ and the maximum remains at that wavelength to the greatest depth reached of 12 m. The upwelling radiation was measured immediately afterwards by placing the photometer in the inverted position. The first measurement was made 50 cm above the surface with successive measurements at depths below the surface to 10 m (Fig. 3). Again the point of maximum transparency moved to 560 mμ.

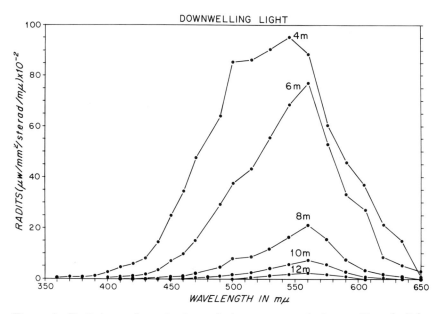

Figure 2. Preliminary determinations of the spectrum of down-welling daylight
   (sun plus skylight) at the indicated depths in Woods Hole Harbor, August 9,
   1967.

A second series of observations was made in Massachusetts Bay
east of Boston Light Ship where the water is more than 100 m deep.
The spectrum of downwelling light was measured at six depths from
5 m to 26 m as shown in Figure 4. The shift in the region of maximum
intensity moved from that characteristic of the surface to a band cen-
tering at about 530 mμ. The spectrum of the upwelling light was recorded at 6 depths from 1 m to 17 m where measurable spectrum was
confined between 500 to 560 mμ (Fig. 5).

At the same time, measurements of upwelling light were made
with the second radiometer from the airplane flying at an altitude of
about 300 m. In order to eliminate as far as possible the light reflected
from the surface itself so as to measure chiefly the radiation which
had been scattered upward from beneath the surface, the following
procedure was adopted: taking advantage of the fact that light reflected from the surface is polarized, the radiometer was directed away
from the sun at Brewster's angle of incidence of 53°. At this angle,
no detectable amount of horizontally polarized light is reflected from
the water's surface. The vertically polarized light, which is reflected from
the surface, was cut out by placing a polarizing filter in a vertical
position over the receiving window of the radiometer. In Figure 6 the
results of the entire operation are summarized. Again, it must be emphasized that the measurements are entirely of a preliminary or ex-

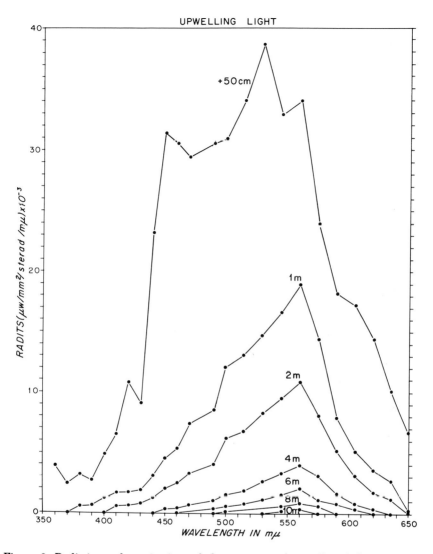

Figure 3. Preliminary determinations of the spectrum of upwelling light at 50 cm above the surface and at the indicated depths in Woods Hole Harbor, August 9, 1967.

ploratory nature. At the bottom, the records of the spectrum of upwelling light made from the ship at depths of 1 and 2 m are repeated. At the top, the record made from the plane with no polarizing filter in place is presented. The third and fourth curves show the effects of placing the polarizing filter first in the horizontal position and then in the vertical position. With the horizontal polarizer, it is seen that a large amount of blue light is recorded and this is interpreted as consisting

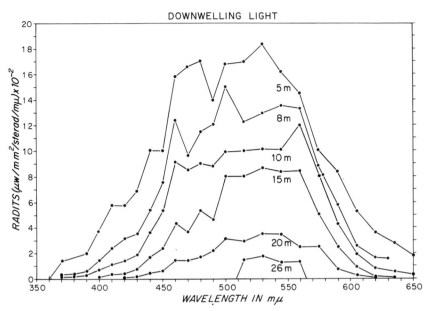

Figure 4. Preliminary determinations of the spectrum of downwelling light at the indicated depths east of Boston Light Ship, September 8, 1967.

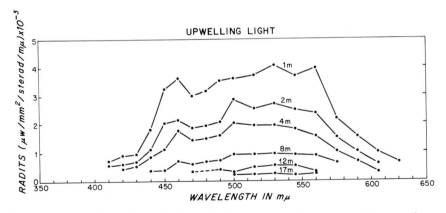

Figure 5. Preliminary determinations of the spectrum of upwelling light at the indicated depths east of Boston Light Ship, September 8, 1967.

chiefly of skylight reflected from the water surface. With the vertical polarizer, a large amount of this blue light is cut out and the curve begins to approach the shape of the record obtained at 1 m. When the values obtained with the vertical and horizontal polarizer are added together, the curve second from the top is obtained. This is similar to, but lower than the top curve, obtained with no polarizing filter. The difference is believed to be caused by the inevitable loss of light in passing through the filters at all wavelengths.

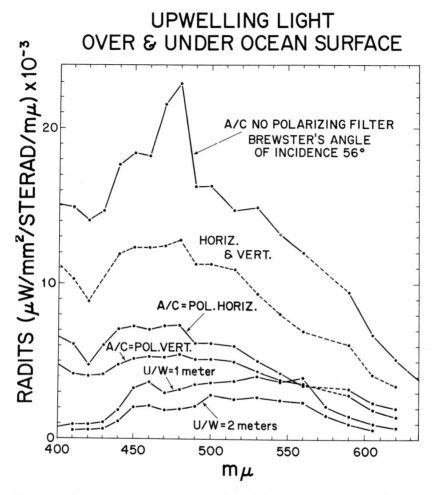

Figure 6. Approximate spectra of upwelling light at a station east of Boston Light Ship. Upper four curves recorded from airplane at altitude of 300 meters. Lower two curves recorded from ship (see text for explanation), September 8, 1967.

These very preliminary results show the feasibility of measuring upwelling light beneath the surface, at the surface, and just above the surface, and comparing the spectra with simultaneous measurements made from an airplane overhead. The curves exhibit reasonable shifts in the spectrum in accordance with the optical properties of the water at the localities visited. No extra absorption is found in the region of the red band chlorophyll since the spectral range of the radiometers used did not extend beyond 650 mμ.

Our plans for the immediate future include the construction of one or more highly sensitive photometers with a range extending to 700 mμ, and with the capability of distinguishing differences between narrow bands of the spectrum. We know that the chlorophyll of living plants, and other pigments of importance in the water, modify the spectrum of light passing down through the water and then scattered up out of the water. Accordingly, if a photometer with adequate characteristics to detect these differences can be built, and if methods can be perfected for eliminating or allowing for the effects of the water surface and of the atmosphere above the water, then meaningful records can be obtained of differences in the spectrum of the water from aircraft or from spacecraft. The procedure would be extremely valuable in permitting the rapid mapping of water masses over large regions of the ocean for the benefit of both physical and biological oceanographers, with the added possibility of finding areas rich in organisms of commercial importance.

SUMMARY

The spectral distribution of visible daylight entering the sea was determined by means of a special, high-sensitivity radiometer at two locations off New England. Both downwelling and upwelling radiation were measured above the sea surface and at a series of depths from shipboard. Upward scattered light was measured simultaneously from an airplane. Since changes in the spectrum are affected by differences in the plankton and/or in the dissolved and particulate matter in the sea, the procedure suggests a method for the rapid delineation of water masses from the air, and possibly of areas of high productivity.

REFERENCES

Clarke, G. L. 1965. Transparency, bioluminescence and plankton. *Oceanography from Space* (G. C. Ewing, ed.) Woods Hole Oceanographic Institution, Ref. 65-10, p. 317-319.

Clarke, G. L. and M. G. Kelly. 1965. Measurements of diurnal changes in bioluminescence from the sea surface to 2000 meters using a new photometric device. Limnol. Oceanogr. 10 (suppl.) :R54-R66.

Clarke, G. L. 1967. Light in the sea. Oceanology International 2:38-47.

Clarke, G. L. and E. J. Denton. 1962. Light and animal life. In *The Sea: Ideas and Observations*, Sect. 4, Ch. 10, Interscience Publishers, Ltd., London. pp. 456-486.

Conrod, A. and R. Kezar. 1966. Final report on spot experiments on sky brightness measurements in visible spectrum for the staff program Vol. II of II, p. 1-57. Unpublished manuscript, Experimental Astronomy Laboratory, Massachusetts Institute of Technology.

Smith, R. C. and J. E. Tyler. 1967. Optical properties of clear natural water. J. Optical Soc. America 57:589-595.

Tyler, J. E. and R. C. Smith. 1967. Spectroradiometric characteristics of natural light under water. J. Optical Soc. America 57:595-601.

12

Aerial Photographic Studies of Shallow Water Benthic Ecology

Mahlon G. Kelly and Alfred Conrod

INTRODUCTION

Aerial photography has proven to be a very productive tool for the terrestrial ecologist and increased use of satellite photography is projected. Surprisingly, however, aerial photography has been used very little in ecological studies of the marine environment. This is particularly surprising since a detailed synoptic understanding of the distribution of bottom biota is very difficult to obtain from the surface of the ocean, and since marine biologists and oceanographers have for some time realized the necessity of gaining a synoptic view of the environment they study. Aerial photography does in fact reveal many biological features and distributions that cannot be recognized from on or below the surface even after considerable field work.

Aerial photography of submarine features was suggested during World War I (Lee 1922) and was used by the Navy for bathymetric reconnaissance during World War II.. Geologists have made some use of aerial photographs of submarine features; for example Cloud (1962) has studied the Bahama Banks and the Island of Saipan and Newell and and co-workers have worked extensively on the Bahama Banks. To our knowledge however, the only biological studies that have been published using aerial photography were those done in a gross survey of the bottom features near St. John, Virgin Islands by Kumpf and Randall (1961).

It may be argued that aerial photography is limited to shallow areas that lie along the coast and by the turbidity of the water in productive areas or areas with much land runoff, but features of 10 or 15 ft. depth may be photographed even in such turbid water. Coastal regions however are the areas of the ocean that are of most immediate importance to man, since they can be most easily utilized and are most heavily polluted by man as he uses the resources in the ocean.

Three types of biological information may be gained from aerial photography: First, with sufficient fieldwork it is possible to map the distribution of the biological communities and to delineate quite subtle changes in the distribution of the plants. We realize the ambiguity of the word community, but for the present purpose it seems clearer than assemblage or association and less ambiguous than the word biotope, which is used differently by different workers. Second, it is possible to identify anomalous features that would not have been recognized using conventional survey methods. Finally, and this is yet to be investigated in detail, it should be possible to learn much of the dynamic interrelationships between the biological communities and the environment by analyzing the distributional patterns and the geographic variations shown in the aerial coverage.

PHOTOGRAPHIC STUDIES OF BOTTOM COMMUNITIES ON THE BAHAMA BANKS

We have examined an area extending about 23 miles south from Bimini on the west edge of the Bahama Banks to find what may be learned using aerial photography. This site has the advantage that good Gemini photography is available, that the water is clear, that weather conditions are usually favorable for field work and aerial photography, that human disturbances are at a minimum, and that good support facilities at the Lerner Marine Laboratory are available. Aerial photography was obtained from NASA aircraft using a Wild RC-8 camera. Photographic coverage was flown at various heights from 2000 to 25,000 ft. using various combinations of filters and films. An uncontrolled mosaic was made of the 25,000 ft. color photography and features seen on the mosaic were examined in detail so as to map the distribution of the plant communities in the area (Fig. 1). The features seen on the mosaic posed some interesting questions and showed important relationships that would not have been recognized without the aerial coverage.

The major organisms in each of the communities were identified and the zonation and distribution of the communities around the cays and rocks, the edge and ecotonal locations between the communities, and other features were examined in detail. Divers were towed over the bottom and meter square quadrats were photographed at regular points along the transects to examine the uniformity of the plant cover.

The major communities and areas of hard and soft bottom are shown in Fig. 1. The hard bottom biota were not examined in sufficient detail to allow photointerpretation and plotting on this map, but the soft bottom communities may be divided into three types, one characterized by a dense cover of the grass *Thalassia testudinum* (Konig), a second characterized by patchy cover of the thin leaved grass *Diplanthera wrightii* Aschers (=*Halodule wrightii* Aschers), and a third without any macroscopic plants that was characterized by well-sorted, rippled

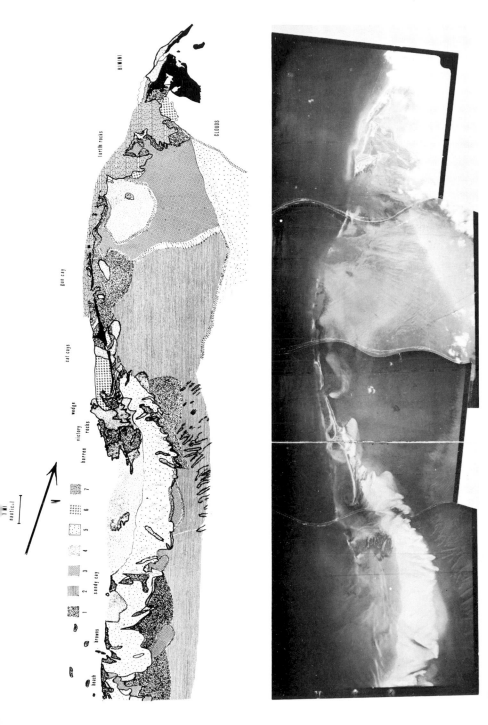

Figure 1. Photomosaic of the Bahamas test site and a vegetation map made from the mosaic and field observations. Photographed from 25,000 ft. with a Wild RC-8 camera and 6″ lens. Map key: 1. Dense *Thalassia* community. 2. Sparse *Thalassia* community. 3. Patchy *Thalassia* community interspersed with the *Diplanthera* community. 4. *Diplanthera* community. 5. Clean sand with no macroscopic vegetation. 6. Rock with patchy sand overlay. 7. Rock bottom.

sand. The *Thalassia* communities may be in turn divided into those with a dense cover of grass with relatively few algae present and those having a less dense, patchy cover but with more numerous algae, *Diplanthera* and *Syrongidium filiforme* Kutz. The *Diplanthera* communities contained numerous patches of algae and the grass was usually distributed in irregular patches 3 to 18 m. in size (see Fig. 2-a). The sandy bottom areas were found in shallow regions with strong current and wave scouring.

The *Thalassia* communities were associated with the algae *Penicillus*, *Rhipocephalus*, *Avurainvillea*, and *Udotea*, and the grasses *Diplanthera* and *Syrongidium* were present in some locations. The *Diplanthera* communities on the other hand included the algal genera *Penicillus*, *Rhipocephalus*, *Avrainvillea*, *Udotea*, *Halimeda*, *Acanthophora*, and two unidentified forms, although no other grasses were found. The distribution of algae in the *Thalassia* beds was fairly uniform, and the algae were sparse, whereas the algae associated with *Diplanthera* were usually found in discrete patches. Since *Thalassia* acts to retain sediment, the sediments associated with it are finer and less well sorted. Numerous animals were found, such as pen shells, burrowing pelecypods, sedentary tube forming polychaetes such as *Arenicola*, a few corals (*Manicina areolata* [Linn.]), and at least four species of echinoids. Because of the wave washed sand the *Diplanthera* communities were characterized by less numerous but more motile animals. Such organisms as burrowing pelecypods, errant polychaetes, and a few bottom fish were found. The distribution of the grass in these communities may be seen in Figs. 1 and 2-a.

The sandy areas consisted of a shoal area southeast of Bimini with well sorted, rippled shell fragments and oolite sand, and an extensive range of sand bars between South Cat Cay and Beach Cay that have been described in detail by Ball (1967) (Fig. 1). Only a few burrowing pelecypods and errant polychaetes were found in the sandy areas. The sand bars south of South Cat Cay consisted of more than 90% oolite sand with a strong white reflectivity obvious in the aerial photographs. The sand in both areas was extensively rippled and the ripples are visible on the photographs.

Although no quantitative measurements were made, it was obvious that slight gradations in the amount of plant cover could be detected from the aerial photographs. This is particularly seen in the area east of Gun Cay, where patchiness is quite conspicuous in the photographs although the patches were barely detectable from the surface or while diving (Figs. 1 and 2-a).

An ecotonal situation was found east of the line between Bimini and Gun Cay (Figs. 1 and 2-a) where variable patches of *Thalassia* grew among the *Diplanthera* community. The grass is seen in corrugated rows that in turn appeared rippled along their length (Figs. 1 and 2-a). It seems likely that this rippling and patchiness is in some way related to the distribution of the generally east-west tidal currents although the exact relationship has yet to be investigated.

An interesting zonation was seen on the eastern edge of the spillover lobes south of South Cat Cay. The *Diplanthera* community was found along the eastern edge of the bars, sandwiched between the sandy area and the *Thalassia* beds as shown in Fig. 1. This is strongly suggestive of the spatial succession of grasses often found on beach dunes. It may be that *Diplanthera* acts to consolidate the sediments to the extent that they can then be colonized by *Thalassia*.

Although the hard bottom biota was not studied as extensively as the soft, certain generalities may be made. The hard bottom usually lay to the west of the Cays and may be distinguished in the photographs by its texture in relation to the sandy bottom. Because the hard bottom sloped away from the banks, the algae were zoned with depth as is seen in the aerial photography. There was, however, great variation in zonation from place to place and thus the zonation is very difficult to generalize. Several different types of bottom cover were observed but the species composition of each type varied and these types were not considered to be discrete communities or associations. By far the most widespread type of cover was a low matted assemblage of finely branched calcareous algae that densely covered the rock. This can be distinguished by its relatively light tone. A second type of cover was characterized by frondose algae, mainly *Sargassum* and *Dictyota* and this showed on the photographs as a conspicuously zonated dark brown tone (Figs. 1 and 2-b). A third type of bottom cover consisted of multi-branched alcyonarians of 5 or 6 different species. These generally grew where the rock was covered by a thin layer of sand and may be distinguished on the photographs by a lack of texture and darker tones. The sand presumably prevented the growth of algae; if the sand was washed aside, dead matted algae were often found on the rock beneath. These three types of cover however should not be generalized to all of the rock bottom. The rocky bottom communities in some areas were very complex, especially where the surface was incised or rough. The different types of hard bottom cover are visible on the aerial photography and it should be possible to describe and map the communities that have been found there, as it has been for the sandy bottom communities.

AERIAL DETECTION OF MAJOR BOTTOM FEATURES

Several features were found on the aerial photographs that could be distinguished from the surface of the water only after they were seen in the aerial photography. This was either because the features were too large to be resolved from the surface or because the differences in tone were too slight to be seen from the surface or while diving. There were many features of this sort; the most interesting and conspicuous are described here.

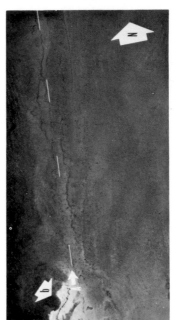

Figure 2-b. 5,000 ft. photograph of the area immediately south of Beach Cay. a. The "ray," on the bottom formed by a lack of frondose algae. b. The relatively shallow (8 ft. at low tide) *Sargassum* zone on the rocks extending south from Beach Cay. c. Lunate sand blowouts in the *Thalassia* beds.

Figure 2-c. 5,000 ft. photograph of an area immediately north of Turtle Rocks. a. The line of incised bottom grooves extending between the cays. b. Turtle rocks.

Figure 2-a. 5,000 ft. photograph of an area west of the rocks north of Gun Cay. Small arrows point to emergent rocks. a. Dark patches of *Thalassia* interspersed in the *Diplanthera* community. b. Corrugated rows and ripples of *Thalassia* interspersed in the *Diplanthera* community. c. Lunate sand blowouts or cusps in the *Thalassia* beds. d. Edge of the dense *Thalassia* beds abutting onto the hard rock bottom.

The most common of such features were grass-free cusps 25 to 30 m. wide found along the edges of the banks in the dense *Thalassia* beds (Figs. 1 and 3). The cusps were generally half-moon-shaped although two or three often coalesced. The grass at the convex edges of the cusps was generally undercut with a 1 m. scarp with *Thalassia* rhizomes hanging into the undercut area (Fig. 3). The bottom usually consisted of coarse, well-sorted gravel with sparse *Halimeda* and *Udotea* growing on it, although the bottom of the illustrated cusp was hard rock. The concave edge consisted of fairly level diffuse grass, either *Thalassia* or *Diplanthera*, and was not sharply demarcated, as seen in the photographs. The convex edge generally faced the west, in the direction of the oncoming surf. The cusps were probably formed during high-energy storms by erosions and uprooting (Ball 1967). The grass may recolonize the level, leeward side of the cusp, thus forming the diffuse edge. Some cusps were found that had been nearly grown over, although the scarp was still visible while diving. The formation of these cusps probably accounts for some of the large volumes of *Thalassia* found along beaches after hurricanes (Ball 1967); Thomas *et al.* 1961). The cusps seem to indicate areas of maximum wave action and erosion, and this in turn is important in determining the distribution of the *Thalassia* beds.

Another feature seen in the aerial photography (Figs. 1 and 2-a) is the regular distribution of the grass patches in some areas. Usually the patches were round, 6 to 18 m. in diameter, and grew to the east of a major *Thalassia* bed. The westward or surfward edges of the major beds were usually sharply defined. Often, as in Fig. 2-a, the patches coalesce to form fairly regular ripples and lines in the direction of the tidal currents. These grass distributions are probably related in some way to the distribution of the east-west tidal currents in the area, and should be important to an understanding of the ecology of the grasses. Extensive current measurements will be necessary before the patchiness and distribution may be explained.

Several features were seen on the hard bottom that seem quite difficult to explain. An example of these is shown in Fig. 2-b. It was at first thought to be a "ray" of sediment extending to the west of the small rocks, but on examination was seen to represent the distribution of two types of hard bottom communities. The light part or "ray" consisted of white, matted, jointed calcareous algae of several species. The dark pattern to the north and south represented frondose algae, mostly of the genera *Dictyota* and *Sargassum;* matted algae grew below the frondose types. No immediate explanation is apparent for this regularly shaped feature, although it may be related to wave reflection from the rock. Similar features were seen at several points where isolated rocks occurred.

Another surprising feature of hard bottom was the parallel incised grooves extending between several of the cays in a north-south direction as shown in Fig. 2-c. The grooves were 3 to 10 m. wide, separated by

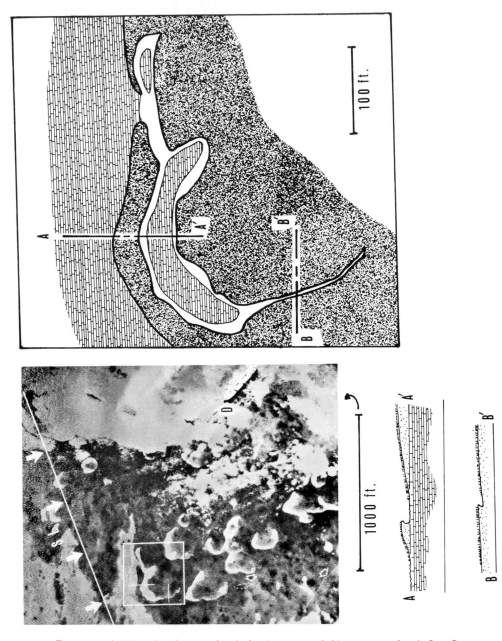

Figure 3. A 2,000 ft. photograph of the lunate sand blowouts south of Cat Cay. D. A ship dredging the oolite sand. SL. The surf-line, showing the direction of the approaching surf. Note that the blowouts are oriented with the surf-line. The small arrows point to lines in the grass beds where boats have run their keels aground (depth about 15 ft). The position of the inset drawing is shown on the photograph. The drawings show details of a blowout as determined by diving and inference from the photograph. Section A-A' shows the rock bottom and relative surface contours; section B-B' shows the contours where grass has grown back into the blowout (vertical scale exagerated). Key to the symbols used is given with Fig. 1.

8 to 20 m. of flat wave-cut rock, and the bottoms of the grooves were covered with coarse, well-sorted, shell-fragment sand. Presumably these grooves represent areas where softer limestone has been removed by solution and abrasion.

Other features were found that were difficult to distinguish from the surface. Some of these were the conspicuous patterns and positions of the algal communities on the hard rocks, the distribution of the margins of the grass beds and the exact positions of the beds in relation to the cays and the hard rock, the location of offshore rock bars, the depth zonation of the algae, the rippled patterns in the sandy areas, and the above mentioned zonation of grasses at the edges of the sand bars. It will be interesting to compare some of these features with those in similar environments on the other side of the Florida Straits and elsewhere in the Bahamas, for if they are fairly common they may offer important clues as to the ecology of such areas. It has become apparent that aerial photography shows distributional patterns and bottom features that cannot be distinguished from the surface or underwater. Such features certainly would have been noticed and thoroughly investigated were they found on land, but since the sea surface has remained a barrier to synoptic investigation their explanation remains a matter for the future.

SATELLITE PHOTOGRAPHY

It should be possible to gain considerable information about the distribution of shallow water bottom communities with photography from space, which would have the advantage of covering very large areas. Figure 4 shows an enlarged portion of a transparency taken from 100 miles altitude during Gemini XII. The accompanying vegetation map shows that the locations of the various communities can be seen even in such low resolution photography. It is apparent on examination of the Gemini photography that there are many regions of the world where detail can be seen on the bottom, and if the resolution of the photography is increased in the future it should be quite possible to perceive good detail and to obtain considerable information on plant community distribution. It should be obvious, however, that field work will continue to be necessary in order to properly interpret the features that are seen in either aerial or satellite photography.

DISCUSSION

A test site was examined on the western edge of the Bahama Banks, and the major soft-bottom communities were described and mapped using aerial photography. A number of interesting distributional features and anomalous conditions of the bottom biology and geology were seen in the aerial photography that could not be detected from the surface or below. The entire field survey has so far required 10 days of diving and sampling in the field. It appears that a very great saving in time

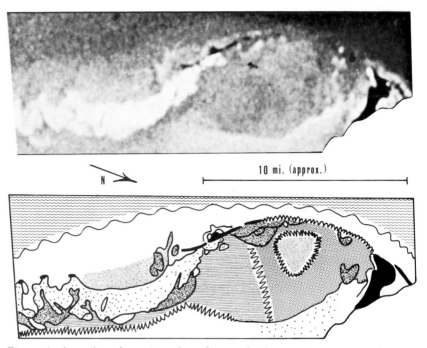

Figure 4. An enlarged portion of a photograph showing the test site from 120 nautical miles altitude during the flight of Gemini XII. The accompanying vegetation map was plotted using the features that may be seen on the photograph. The western wavy line represents the extent of visible bottom. The islands are identified and a key is given with Fig. 1. Note that relatively fine differences in grey tone may be used to infer the position of bottom communities.

may be made by using aerial photography to survey the locations of bottom communities. A similar study was made by Newell and co-workers describing the geological facies of this area (Newell, *et al.*, 1959), but they mainly used conventional sampling and observing techniques with a limited amount of aerial observation. Their survey required a period of several years, and although they described the geological features in much greater detail and over a wider area, it still appears that considerable time saving may ensue from the systematic use of aerial photography. It is apparent that aerial photography allows a much more precise determination of the location of features and bottom communities and allows distributional features, gradations in the plant cover, and anomalous conditions to be detected that could not be found using conventional techniques.

During our work it became apparent that high altitude coverage was much more satisfactory than it is in terrestrial work of similar nature. This is because the communities and large scale features are not here identified by discrete objects such as trees, streams, etc. that need to be finely resolved on the film, but are characterized by the tone and texture of the image. It is often necessary to check certain detailed features using lower altitude coverage but such photographs could be taken from light planes using hand held cameras and need not be a part of the original photographic coverage. The most useful photographic coverage was the 25,000 ft. material used in the field studies, because it enabled orientation of the features in the field without using an unmanageable number of pictures. It seems likely that good satellite photography would also be very useful.

A final important point must be made: It is not possible to undertake investigations of this sort without considerable "ground truth analysis," and such field work cannot be instrumented as in studies of meteorology or physical oceanography. Although it is not possible to instrument the field study, continuous measurement throughout a period of time is not necessary and it should be no more difficult to conduct such field studies than to design and use the instrumentation necessary to measure equivalent meteorological or physical parameters. In order for aerial and satellite coverage to be used, however, marine ecologists will have to become aware of the advantages in the use of these techniques in their studies. Subjectively we have found that aerial photography gave a perspective on the distribution of marine bottom communities that could be obtained in no other way, and we feel that these techniques should help biologists working in coastal areas to look at some of the large scale community distributions that are of importance to our knowledge of the seas along our coasts and estuaries.

SUMMARY

A test site on the west edge of the Bahama Banks has been examined using aerial and satellite photographs. The location, depth, and qualitative characteristics of the bottom biota, sediments, and morphology can be determined. The density of the bottom algae and "grasses" as well as the dominant species can be estimated, and many of the relationships between different bottom communities and the environment can be qualitatively estimated. The work reported is preliminary, but the method shows promise as a valuable tool in shallow water ecological studies.

Acknowledgements

This work was supported by Naval Oceanographic Office contract no. N62306-2082 to the Experimental Astronomy Laboratory of the Massachusetts Institute of Technology. The field work was supported in part by the Lerner Marine Laboratory of the American Museum of Natural History and Office of Naval Research contract 552 (07).

REFERENCES

Ball, M. M., E. A. Shinn, and K. W. Stockman. 1967. The geologic effects of Hurricane Donna in South Florida. J. Geology, 75:583-597.
Cloud, P. E. 1962. Environment of calcium carbonate deposition west of

Andros Island, Bahamas. U. S. Dept. of the Interior, Geological Professional Paper, 350:1-138.

Kumpf, H. E., and Helen A. Randall. 1961. Charting the marine environments of St. John, U. S. Virgin Islands. Bull. Marine Sci., 11:543-551.

Lee, W. T. 1922. The face of the earth as seen from the air. Amer. Geographical Society, N. Y., 110 pp.

Newell, N. D., J. Inbrie, E. G. Purdy, and D. L. Thurber. 1959. Organism communities and bottom facies, Great Bahama Bank. Bull Amer. Nat. Hist., 117:181-228.

Thomas, L. P., D. R. Moore, and R. C. Work. 1961. Effects of Hurricane Donna on the Turtle grass beds of Biscayne Bay, Florida. Bull. Marine Sci., 11:191-197.

Surface Temperature Patterns of Hudson Bay from Aerial Infrared Surveys

Wayne M. Wendland and Reid A. Bryson

INTRODUCTION

Hudson Bay, the large North American inland sea of some 520,000 km², exerts a significant influence on the air overhead and on its surrounding shores. During winter, the ice cover creates a surface much like the nearby tundra. As air passes over the open Bay in summer, its temperature and humidity may be markedly changed. These induced changes to the characteristics of the air can be large, as seen from previous work (e.g., Bryson and Kuhn 1962; Lamont 1949; Hare and Montgomery 1949; Burbridge 1951). If the modification process is to be completely understood and predicted, the surface temperature structures of the Bay must be known.

Since Hudson Bay is a regional source of air masses passing over eastern Canada (Bryson 1966) and modifies all air which passes over it, changes in the characteristics or frequency of Hudson Bay air are important to the climatologist, and those interested in the ecology of the area. Until recently, surface temperature information for Hudson Bay primarily consisted of a few isolated measurements. A review of the early (pre-1950) work can be found in Hachey (1933) and Barber (1967). The first surface thermal analysis of the Bay using more than just a very few data points was reported by Barber and Glennie (1964). Their report summarizes the oceanographic survey of the "Theta" and "Calanus" vessels during 1961. Although data are presented for most of the Bay for August, the data was gathered over a four week period. Do these charts closely approximate the sea surface temperature (SST) distribution at a given time (implying little time change to the Bay's surface temperature patterns), or do the reported patterns represent a smooth mean, obscuring changes in space and time?

In 1965 and 1967, the SST was remotely sensed with an airborne infrared thermometer. In 1965, a Barnes model IT-2 infrared thermometer was used and in 1967, the Barnes model 14-320. Both instruments are sensitive to radiation emanating from a circle of about 40 ft. diameter when viewed from 1000 ft. altitude. The aircraft flew a predetermined course, and data were gathered continuously when there was clear air between the plane and the water. Each of the surveys was the result of data gathered over about 3800 nautical miles of track.

INTERPRETATION OF DATA

The radiation temperature of the bolometer's "target" can be determined with the Stefan-Boltzman relationship, if the emissivity of the radiator is known. The determined radiation temperature need not necessarily equal the SST as measured by a ship. Infrared radiation emanates from the upper few microns of the sea, and measurements by ship usually involve a bucket and thermometer, or perhaps a calibrated sensor suspended in the upper layer of water. The radiation temperature is a true "skin" temperature, and is dependent upon solar heating, mixing of the surface water layer, and evaporation. These processes can cause the "skin" temperature to differ from a conventional bucket temperature by a few degrees. The data presented on the accompanying Figures represent radiation temperature.

The IR bolometer receives radiation between 8 and 13 microns. Although this approximates the atmospheric "window," the atmosphere is not completely clear, i.e., it absorbs some radiation from the sea surface and re-radiates it at its own (air) temperature. Without a temperature-humidity profile beneath the aircraft, the effects of such spurious radiation cannot be known, and error is introduced. Lenschow and Dutton (1964) have shown that flight altitudes less than about 300 meters essentially eliminate the effect of atmospheric absorption and reradiation. Osborne (1964) claims that arctic air in summer should not substantially alter the radiation emitted from the water surface due to the temperature and moisture characteristics of arctic air. During our surveys, flight altitude temperatures were within 4 C of the water surface temperatures, suggesting no substantial reradiation error. In July 1967, the air temperature was consistently ten to twelve degrees warmer than the water surface. This suggests that the July data might have been altered by atmospheric reradiation, but probably not to a significant extent. The bolometer was calibrated before and after each flight.

Isotherm configuration was independently assessed by using SST records solicited from freighters and supply ships operating in Hudson Bay during the time of the surveys of August and October 1967. The absolute values of the ship data cannot be used to calibrate the airborne bolometer since the times of the two measurements differed by as much as a few days, and because ship sensing devices may not be themselves calibrated. Nonetheless, the large scale isotherm configuration from the ship data should be representative of the surface patterns.

July 1967 Survey

Figure 1 presents the temperature pattern for 12 and 13 July 1967. About two-thirds of the Bay was covered with seven- to nine-tenths ice. The open area in the northwestern Bay was already rather warm. Even near the periphery of the ice, the water temperature was well above the melting point. The warmest water surveyed was just north of Churchill (12.4 C), with a warm tongue extending northeastward toward Southampton Island. A cold tongue reached southward just off the west coast of the Bay. Air moving across the Bay need not leave its surface as a homogeneous airmass, since the melting ice surface would be near zero degrees Centigrade, and the open water significantly warmer.

August 1967 Survey

The only ice seen at this time (26, 27, and 28 August 1967) was located in the extreme northeastern part of the Bay. The last remnants of ice cover disappeared from Hudson Bay during the week prior to the survey.

The warmest water was found in the western Bay (Figure 2), although the temperatures were below those in July. A warm tongue was found reaching northeastward from Churchill, and the cold tongue was also found along the west coast from near Chesterfield Inlet. The area where the last ice had been centered (about 58 N, 82 W) was marked with a cold pool with temperatures of 3.5 C or less. The area about Mansel and Coats Islands in the northeast was coldest of the surveyed area. Since pack ice still existed in this far-north area (long past the time of strongly positive net radiation), the cold water was not surprising. Figure 3 presents the only data obtained during August 1965. This survey (8 August 1965) falls about midway in the course of the season between the July and August surveys of 1967. The isotherms must be tentative due to the few data used, but the temperatures are similar to those of 1967.

September 1965 Survey

A cold tongue was again found (22, 23 and 24 September 1965) parallel to the west coast, and warm water reached northeastward from Churchill (see Figure 4). The largest feature of the temperature pattern was the cold pool located in the eastern half of Hudson Bay. We postulate this to be the melt water of the last ice. The melt water is colder than surrounding water, but it is less dense due to its low salinity, and therefore remains as a surface feature, long after the ice disappears.

October 1967 Survey

The patterns of this survey (3, 5, 6 and 7 October 1967) were decidedly different from the earlier patterns (see Figure 5). The temperature gradient is generally directed to the north, whereas earlier it had been directed to the east or northeast. Intense cooling of the surface waters obviously had been occurring. A warm tongue from Churchill

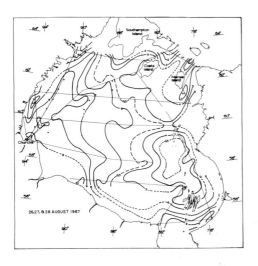

Figure 1. Surface temperature pattern of 12 & 13 July 1967.

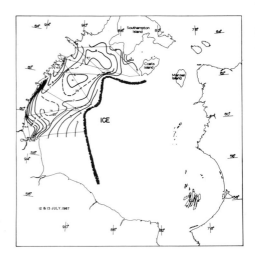

Figure 2. Surface temperature pattern of 26, 27 & 28 August 1967.

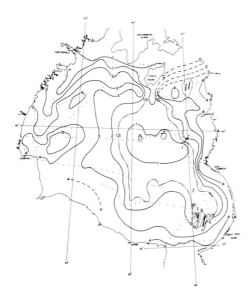

Figure 3. Surfacing temperatures of 8 August 1965.

Figure 4. Surface temperature pattern of 22, 23 & 24 September 1965.

reached northeastward, but it was not as well defined as earlier. A suggestion of the cold tongue was found along the west coast. The warmest water found was near the Belcher Islands, and temperatures in this area were not substantially different from August.

DISCUSSION

Barber and Glennie's chart of the August patterns of 1961 are quite similar to our August survey of 1967. The major difference is found in the eastern Bay. No cold pool was observed in 1961. This is probably because the last ice of 1961 was located in the southwest Bay and not in the east as in 1967.

The first open water (that of July and August) warmed rather rapidly with time and in close proximity to the existing ice. Figure 6 presents three examples from 28 August 1967. Why, then, did not the last melt water (that of September) also warm, instead of remaining a surface cold feature? This is presumably the result of timing. The early open water rapidly warms because net radiation is strongly positive in June and July. By late August or early September, the net radiation has already decreased to between 25% to 50% of its maximum annual value. The above demonstrates the heavy energy drain imposed on the Bay by the persistence of the annual ice cover. Hudson Bay does not lose the last of its ice until the water has already entered its annual cooling cycle.

The similar results of the month-long August 1961 ship survey and the three-day survey of 1967 suggest that the surface patterns do not change shape rapidly. They are probably not wholly controlled by bottom topography (Wendland and Bryson 1966).

The areal mean surface temperature of Hudson Bay was calculated and the results are shown in Table 1 and Figure 7. Ice was observed during the July 1967 and the August 1961 surveys. To determine the areal mean temperature for the entire surface area at those times, the water temperature under the water-ice interface was assumed -1.7 C (salinity of 30 0/00). The process of ice formation or ablation determines the temperature at the interface.

Break-up begins in June and the Bay is usually ice-free by the end of August. Freeze-up begins in October and results in nearly continuous ice cover by late December. The trend curve in Figure 6 indicates that the mean temperature rises rapidly during the season of strong positive net radiation, and cools rather slowly thereafter. A mean maximum temperature of about 6.2 C is suggested. Because Hudson Bay is so intensely stratified, at least in the upper 25 meters in September (Collin 1966), it is suggested that the curve presented in Figure 7 also represents a trend curve for the heat content of the Bay during ice-free season.

The data from August 1961 on Figure 7 does not appear to be representative of the same population. This is probably because the data were gathered over a 31 day period. The ice which was present at the

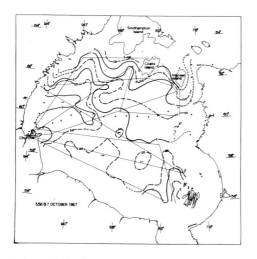

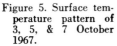

Figure 5. Surface tem-
perature pattern of
3, 5, & 7 October
1967.

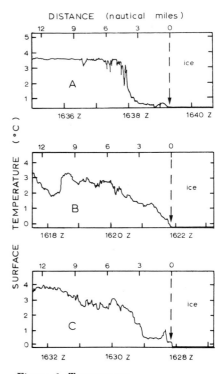

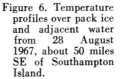

Figure 6. Temperature
profiles over pack ice
and adjacent water
from 28 August
1967, about 50 miles
SE of Southampton
Island.

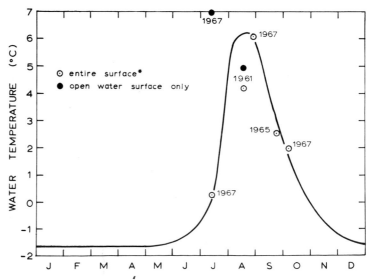

Figure 7. Trend of
areal mean surface
temperature of Hud-
son Bay. See text
for discussion.

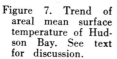

beginning of the sampling period soon melted, the water warmed, and the areal mean temperature would have been seriously altered. Too, the trend curve of Figure 7 can only be considered an estimate based on so few data. Large water bodies sometimes exhibit anomalous surface temperatures of a few degrees magnitude. For this reason, the curve of any particular year may be shifted horizontally or vertically from that in Figure 7, depending on meteorologic and oceanographic forcing functions. The solid dots of Figure 7 represent the ice-free areal mean surface water temperature. These values must be greater than the areal temperature of the entire Bay, if partially ice covered, and must be equal if the entire Bay is open.

Table 1. Trend of the mean areal surface temperature of Hudson Bay.

DATE	TEMPERATURE (degrees Celcius)	
	Ice-Free Area	Entire Surface Area
12-13 Jul 1967	7.00	0.26
3 Aug to 3 Sep 1961	5.03	4.23
26 to 28 Aug 1967	6.11	6.11
22 to 24 Sep 1965	2.57	2.57
3 to 7 Oct 1967	2.02	2.02

CONCLUSIONS

The temperature surveys of 1961 summarized by Barber and Glennie (1964), and those of 1965 (Wendland and Bryson 1966) and 1967 showed similar large scale patterns on Hudson Bay, at least throughout August, September and perhaps early October. The patterns found on surveys completed during a few days were very similar to those found on surveys which were compiled from a month's record. Where Hudson Bay joins Hudson Straits is probably the region of coldest water throughout the year. The warmest area is generally found in the south and over the shallow water north of Churchill. This warm water extends northeastward toward Southampton Island. Reaching southward from Chesterfield Inlet, a cold tongue was prominent. The surface thermal patterns may reflect surface water movement.

A signature of the last ice of Hudson Bay apparently can be located throughout the remainder of the ice-free season. The melt water, because of its rather low salinity and therefore relatively low density, remains a surface feature. This feature is not destroyed, although it is warmed by the incoming radiation, because of the stratification of the upper layer of the Bay.

The surface water warmed rather rapidly after ice loss. This apparently is the result of absorption of solar radiation.

SUMMARY

Surface temperature patterns of Hudson Bay are presented from four near-synoptic surveys during the ice-free seasons of 1965 and 1967. An airborne infrared radiation thermometer provided essentially continuous data over a known track. Certain large scale surface thermal features persisted throughout most of the ice-free season, from soon after ice-loss until early October, when feature definition gave way to a weak south-north temperature gradient. The melt water of the last significant pack-ice remained a surface thermal feature throughout the summer and into early fall.

Acknowledgements

Support for this research was provided by the Geography Branch, Office of Naval Research, under contract number 1202(07). The aircraft crews from NAS Patuxent River, Maryland, the technicians from the Naval Oceanographic Office, Washington, D. C., and R. Steventon and W. Ahrnsbrak, Meteor. Dept., Univ. of Wis. contributed much to the success of this project. Their assistance is gratefully acknowledged. The original report of this work appeared as two technical reports of the ONR contract (Wendland and Bryson 1966, 1967). 1967).

REFERENCES

Barber, F. G. 1967. A contribution to the oceanography of Hudson Bay. Manuscript Report Series No. 4. Marine Sciences Branch, Dept. of Energy, Mines and Resources, Ottawa.

Barber, F. G. and C. J. Glennie. 1964. On the oceanography of Hudson Bay, an atlas presentation of data obtained in 1961. Manuscript Report Series No. 1. Marine Sciences Branch, Dept. of Mines and Technical Surveys, Ottawa.

Bryson, R. A. and P. M. Kuhn. 1962. Some regional heat budget values for Northern Canada. Geogr. Bull. (Canada) 17:57-66.

Bryson, R. A. 1966. Air masses, streamlines, and the boreal forest. Geogr. Bull. (Canada) 8:228-269.

Burbridge, F. E. 1951. The modification of continental polar air over Hudson Bay. Roy. Met. Soc., Q. J. 77:365-374.

Collin, A.E. 1966. Introduction and oceanography of Hudson Bay and approaches. *In* Fairbridge, R. W., (ed), *Encyclopedia of Oceanography*. Reinhold Publishing Co., New York.

Hachey, H. B. 1933. Biological and oceanographic conditions in Hudson Bay. 6. General hydrography and hydrodynamics of the waters of the Hudson Bay region. Contrib. Can. Biol. Fish. 7:91-118.

Hare, F. K. and M. R. Montgomery. 1949. Ice, open water and winter climate in the eastern Arctic of North America. Part 1. Arctic 2:79-89.

Lamont, A. H. 1949. Ice conditions over Hudson Bay and related weather phenomena. Bull., Amer. Meteor. Soc. 30:288-289.

Lenschow, D. H. and J. A. Dutton. 1964. Surface temperature variations measured from an airplane over several surface types. J. Appl. Meteor. 3:65-69.

Osborne, M. F. M. 1964. The interpretation of infrared radiation from the sea in terms of its boundary layer. Deutsche Hydrographische Zeitschrift 17:115-136.

Wendland, W. M. and R. A. Bryson. 1966. Aerial survey of Hudson Bay surface temperature-1965. Tech. Rep. No. 28. ONR Contract 1202(07). Meteorology Department, University of Wisconsin, Madison.

Wendland, W. M. and R. A. Bryson. 1967. Aerial surveys of Hudson Bay surface temperature-1967. Tech. Rep. No. 36. ONR Contract (1202(07). Meteorology Dept., Univ. Wisconsin, Madison.

How Multispectral Sensing Can Help the Ecologist

Fabian C. Polcyn
Norma A. Spansail
William A. Malida

Since World War II, many techniques for producing imagery at wavelengths outside the photographic region have been developed. Imagery can now be generated over the entire optical range from 0.3 to 14.0 micrometers, thus spanning the ultraviolet, visible, and infrared wavelengths. At microwave (mm and cm) wavelengths, passive and active (radar) systems have been developed to produce imagery under all weather conditions. While much effort has been spent to develop and use the sensing instrumentation, less effort has been directed toward understanding the relationships between the remote sensor outputs, the dynamic parameters of the vegetation and the terrain surfaces and the illuminating conditions available for observations. Furthermore, much of the experimentation has been carried out with sensors and camera systems confined to one or another relatively broad spectral band in or near the visible region, and often there has been relatively little "ground truth" collected at the time the images were made. In addition, interpretation of the imagery is always subject to the limitations of the photographic process and the skill and training of available interpreters.

At the University of Michigan, a unique approach is under development, centered around the use of a truly simultaneous multichannel optical sensor. This sensor detects more of the information about an object's radiation than has been possible previously and permits a new form of interpretation by providing data not available to the human eye with the added potential for automatic discrimination of objects from among a large number of observations based on their spectral properties. The key component of the first prototype multispectral sensor consists of a spectrometer mated to a conventional optical-mechanical scanner. With twelve channels between 0.4 and 1.0 microns, the system obtains data with automatic registration in both time and position at a rate of 50,000 spectra/sec (Polcyn 1967). Additional channels are obtained with detector arrays optically filtered to operate in selected bands between 1 and 5.5 microns and individual detectors at 8-14 microns and between .3 and .4 microns. Imagery can be produced in each individual band or from combinations of bands. Thus, for each instantaneous look of the scanner, a spectrum can be derived and a decision made by an automatic processor as to the identification of the object without using any spatial shape information. The design for a 30 channel system has been completed (Lowe *et al.* 1966), which will provide for perfect registration of all bands between 0.3 and 14.0 microns. Such an instrument would be the first device able to provide the full benefits of multispectral remote sensing to the study of ecological and resource problems. Presently, research is centered around the existing prototype system, i.e., up to eighteen channels (twelve of which observe a given resolution element simultaneously) when two or more scanners are operated simultaneously.

For the ecologist who wants to understand the capability of multispectral sensing, an appreciation of the types of information that this new sensor provides and its problem areas is an essential first step. Consider the sensor-user chain suggested in Table 1. Sensor designs can provide outputs from which quantitative measurements can be

Table 1 SENSOR-USER CHAIN

SENSOR REQUIRE-MENTS	→	SENSOR OUTPUTS	→	DERIVABLES	→	INFERENCES OR DEDUCTIONS	→	PRAMETER OF INTEREST

derived and inferences can be drawn or deductions made to answer specific questions about ecological parameters of interest. Table 2 gives a specific example of the types of information that are potentially derivable from a multispectral sensor to answer questions about plant productivity. The concept of plant productivity must be subdivided into those aspects which are measurable by multispectral techniques. This step is important no matter which phenomenon is under investigation. The important point is that one should not expect something of the sensor which it cannot provide, even though each added sophistication of the sensory outputs (left side of Table 2) provides increasingly useful information. A given set of sensor requirements, e.g., good signal-to-noise ratio, synchronous multichannel registration, voltage references, and radiance or reflectance calibration, will provide a set of sensor outputs which contain information about a specific parameter. The user must bring to the problem an *a priori* knowledge as to the meaning of the parameter in a given environmental context. A good example of this approach occurs when the sensor generates a spectrum which leads to an identification of a plant species which in turn permits an inference about the subsurface conditions.

Table 2. Sensor-User Chain to Provide Information Concerning Plant Productivity.

SENSOR REQUIREMENTS	SENSOR OUTPUTS	POTENTIAL DERIVABLES	POSSIBLE INFERENCES OR DEDUCTIONS
Single Channel Data →	Voltage →	Spatial Pattern, Time Variation of Relative Image Contrast, Relative Apparent Temperature (IR) →	Plant Community Structure and Stratification, Maturation or Wilting, Effects Due to Change in Energy Balance
Synchronous Multi-channel Data and Reference →	Raw Spectrum (voltage) →	Species Identification if One Has Training Set, Soil Type Identification if One Has Training Set, Enhanced Image Contrast →	Distribution of Plant Species over a Site, Probable Soil Fertility Levels, Change in Species: Maturity, Disease, Nutrient Status, or Moisture Status
Calibration →	Spectral Radiance watt/cm²/ster. →	Outgoing Energy over Large Areas, Spectral Shape, Day to Day Basis →	Energy Budget Calculations, Distribution of Plant Species or Soil Types, Day to Day Basis
Ground References (or Sun Sensor) →	Percent Reflectance →	Absolute Apparent Temperature (IR), Moisture Content of Soil, Species Identification Using Signature Library →	Quantative Knowledge of Thermal Processes, Possible Moisture Availability for Plant Growth, Distribution of Plant Species or Soil Types, over a Region with Minimum Ground Truth

Unfortunately, much of the research to date has been limited to the collection of multispectral data with little collection of the "ground truth" information necessary to properly interpret the various contrasts observed in each channel. It is hoped that in future research programs the necessary ground work will be completed and that much progress will be realized in obtaining a fuller understanding of the potential of the multispectral approach.

APPLICATIONS OF MULTISPECTRAL SENSING TECHNIQUES TO ECOLOGICAL INVESTIGATIONS

A better understanding of multispectral sensor capabilities in relation to the objectives of ecological research may be gained by consideration of the following areas: (1) how multispectral signatures are used to provide information concerning objects or processes of interest; (2) the characteristic features of multispectral outputs; (3) the advantages of using multichannel scanners over photographic systems; and (4) the sources of variation in scanner video data which influence its interpretability.

All objects within a natural landscape absorb, transmit, reflect, and emit radiation selectively. The radiation collected by the optical-mechanical scanners is converted to electrical signals which are recorded on magnetic tape. The characteristics of the scanners in operation at Michigan have been described in detail (Lowe *et al*, 1966). The taped video data may be converted from magnetic tape to one or more of several types of signature data or imagery displays. The pattern of voltage signals, or image grey tone levels obtained for a given object at a given time in each of the different spectral channels represents the spectral characteristics of the object and is called its multispectral signature. Multispectral signatures are used to differentiate objects of interest and to provide information concerning the energy budget of the objects.

An example of how grey tone levels from a film strip display of several channels of data may be used to differentiate objects of interest is given in Figure 1 a, b, and c. Figure 1a is a photo mosaic of an agricultural area near Davis, California. Note that the grey tones registered for fields of rice and safflower are all similar. Now consider Figure 1b, and 1c which are displays of scanner imagery of the same scene in 18 wavelength bands from 0.32-0.38μ to 8.0-14.0μ. By comparing the grey tone levels in the different images for pairs of fields, it is possible to differentiate each field type.

The sensor outputs from a multispectral system can be tape recorded and a number of display formats can be obtained for analysis. Among them are (1) multispectral signatures as a set of voltage levels in each channel, spectral radiance graphs, and percent reflectance as a function of wavelength, (2) in each channel, a voltage vs. time presentation for point to point comparison, (3) imagery in the form of a "moving win-

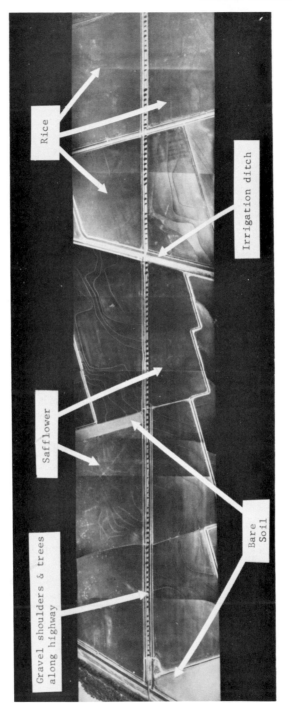

Figure 1a. Panchromatic Mosaic of Davis, California Agricultural Area. 5/26/66; 1600 hrs.; altitude 2000'; sky condition, clear and bright, 10% cloud cover at 30,000 feet; surface temperature 27°C.

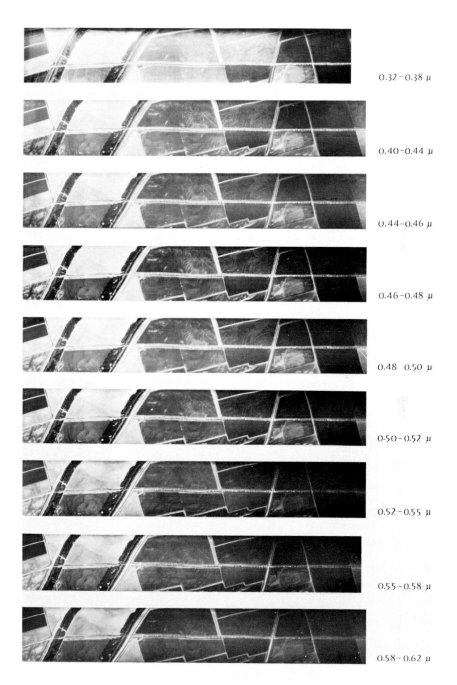

0.32 – 0.38 μ

0.40 – 0.44 μ

0.44 – 0.46 μ

0.46 – 0.48 μ

0.48   0.50 μ

0.50 – 0.52 μ

0.52 – 0.55 μ

0.55 – 0.58 μ

0.58 – 0.62 μ

Figure 1b. Multispectral Imagery of Davis, California Agricultural Area. 5/26/66;
1600 hrs.; altitude 2000'; sky condition, clear and bright, 10% cloud.

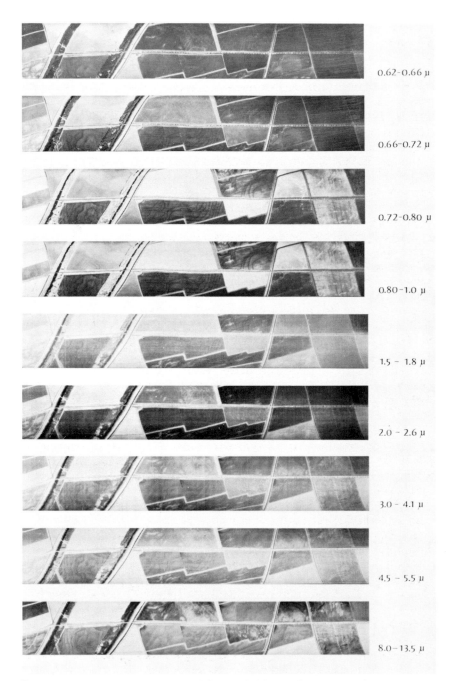

0.62–0.66 μ

0.66–0.72 μ

0.72–0.80 μ

0.80–1.0 μ

1.5 – 1.8 μ

2.0 – 2.6 μ

3.0 – 4.1 μ

4.5 – 5.5 μ

8.0–13.5 μ

Figure 1c. Multispectral Imagery of Davis, California Agricultural Area. 5/26/66; 1600 hrs.; altitude 2000'; sky condition, clear and bright 10% cloud cover at 30,000 feet; surface.

dow" on a C-scope or converted to density variations on film. The main types of image outputs are: (1) standard imagery in each spectral channel; (2) single-channel data that have been processed in some way before being put in the form of an image (e.g., quantization and contouring); (3) single images with contrasts that have been enhanced by combining data in two or more channels by electronic means or by simultaneous display of these data (e.g., a color display); and (4) decision maps that present the results of automatic recognition or classification by computing devices and provide area estimates of particular species. For research purposes, the ecologist will likely want to study all types of image outputs plus calibrated statistical multispectral data from areas of interest, whereas once an operational system has been devised, he may be satisfied with only decision maps and related computations.

A very desirable feature of data obtained from synchronous multichannel, optical mechanical scanners is its amenability to automatic processing. For example, selected spectral channels of the data illustrated in Figure 1b and c have been processed electronically to achieve automatic recognition and mapping of the safflower, bare soil and rice fields (Figure 1). Flexibility of signal processing also permits the removal of the backscatter component of the signals resulting in higher contrast imagery, and the optimization of the signals to reduce the effects of cloud shadows. Other advantages of the multichannel scanner are: (1) the extended spectral coverage which makes possible the establishment of a more unique signature for objects of interest; (2) use of thermal and reflective reference standards permitting calibration of the signals; (3) provision for extrapolation of detailed reflective and thermal measurements (made at a small site) to surrounding areas; and (4) the increased altitude of observation made possible by the greater dependence upon spectral information rather than spatial information. These are important advantages of the multichannel scanner over a multilens camera or several singlelens cameras with multilayer films such as conventional color film and Ektachrome infrared film.

Before discussing major types of contributions of multispectral sensing to ecological investigations and presenting detailed examples of processing that can be performed on multispectral data, it is of interest to consider the sources of variation in the signals which produce the image tones and affect the interpretability of the data. A summary of the sources of variation in multispectral signatures of vegetation is presented in Table 3. There are two types of radiation that reach the sensor from the vegetation, reflected radiation and emitted radiation. The reflected radiation (from 0.3 to about 4μ), originates at the sun, is modified in traversing the atmosphere path, and is reflected by the plant into the hemisphere around the plant according to a reflectance distribution function that depends on the physical properties of the plant. Certain wavelengths of the light are reflected more than others, thereby pro-

Table 3 SOURCES OF VARIATION IN MULTISPECTRAL SIGNATURES OF VEGETATION

ILLUMINATION CONDITIONS	REFLECTIVE AND EMISSIVE PROPERTIES	ATMOSPHERIC CONDITIONS	MULTICHANNEL SENSOR PARAMETERS
Illumination Geometry (Sun angle, cloud distribution)	Spatial Properties (geometrical form, density of plants, and pattern of distribution)	Water Vapor, Aerosols, Etc (absorption, scatering, emission)	Electronic Noise, Drift, Gain Change
Spectral Distribution of Radiation	Spectral Properties (e.g., reflectance or color)		Accuracy and Precision of Measurements On Calibration References and Standards
	Thermal Properties (emittance and temperature)		Differences in spectral responses of systems
SITE ENVIRONMENTAL CONDITIONS	PLANT CONDITIONS	VIEWING CONDITIONS	
Meteoroligic	Maturity	Observation Geometry	
Micrometeorologic	Variety	(scan angle, heading relative to sun) Altitude	
Hydrologic	Physiological Condition	Time of Observation	
Edaphic	Turgidity		
Geomorphologic	Nutrient Levels		
	Disease		
	Heat Exchange Processes		

ducing the color of the object in the visible portion of the spectrum (0.4 to 0.7μ) and producing the spectral differences between various materials; differences on which multispectral discrimination is based.

The illumination from the sun depends on both the time of day and the time of year, as well as on any clouds or atmospheric particles that may act as intermediate reflectors or absorbers. The reflectance distribution function of the plant is determined by its geometry which in turn is dependent upon species maturity, and physiological factors. The plant's condition depends in part on the environmental conditions of the habitat. For most plants, their density and distribution on the ground also affects the radiation received by the sensor. These parameters may change from site to site, from day to day, or even from hour to hour. The constituents of the atmosphere can also change hourly or daily and may scatter extraneous sunlight into the sensor's field of view in addition to selectively attenuating radiation from the plant.

The observation geometry is important because vegetation reflects different amounts of radiation in various directions e.g., it tends to reflect more radiation back toward the source than away from it. These effects have been observed in photographic data (Steiner and Haefner 1965) and multichannel scanner data (Malila 1968). The time of observation and the flight direction determine the relationship of the line of scan to the direction of the sun's angle of incidence. The flight altitude affects the amount of atmosphere that must be traversed by the radiation and also affects the size of the resolved area. The electronics and tape recorder of a multichannel scanner introduce noise, drift, and gain changes which cause additional signal variations. When the data are calibrated there is always some uncertainty about the precise values of the calibration references and standards. This is particularly true when large reflectance standards are used on the ground. Finally, if more than one sensing instrument is used, there are usually differences in their spectral responses which can produce different output values for the same input radiation because the sensor performs a weighted integration of incident spectral power.

The process for emitted radiation (.4 to 14μ) is somewhat different than that for reflected radiation since the plant itself is the main source of radiation. The spectral emittance and temperature of the plant depend entirely on the condition of the plant and the environmental conditions. Since thermal changes occur slowly, the prior thermal history has a strong influence on the temperature of a material at any given time. Thermal data are important, since diseases and abnormalities in plants may be manifested by changes in their heat-exchange processes before their reflectance properties are affected.

Two primary objectives of current ecological research are to understand the basis for biological productivity of terrestrial and acquatic ecosystems, and to develop resource management practices to maintain or increase the productivity and quality of ecosystems. Some of the types of ecological research activities concerned with achieving these objectives

which may be investigated by multispectral sensing techniques include:
(1) the inventory of biotic and abiotic components of natural and modified ecosystems; (2) the assessment of spatial and temporal variation
of selected biotic and abiotic components of ecosystems and (3) the
description of physical and biotic factors controlling biological productivity.

Investigations of the three types given above are currently in
progress or have recently been completed at Michigan's Infrared and
Optical Sensor Laboratory in cooperation with various users. The use of
multispectral techniques for the inventory of biotic components of ecosystems includes the automatic recognition and mapping of plant species
and plant communities. Fields of rice, safflower and bare soil have
been differentiated from one another and from other field types by
their spectral (color) and spatial (gross crop geometry) differences,
and their distributions have been mapped using automatic signal processing (Figure 2). The automatic signal processing techniques used
to map these field types are discussed later. Data has been collected
for plant communities of saw-grass, willow brush, Australian pine, coco-
plum, cattail, and submerged algal mats within the Everglades National
Park and studies will be conducted to determine the feasibility of mapping
these communities by automatic processing techniques. The information
concerning rates and directions of plant succession derived from this
type of inventory should make it especially useful for management purposes in areas of rapid ecological change.

Investigations using multispectral techniques for inventory of
abiotic components include the automatic recognition and mapping of
rock outcrops and surface water bodies. Lava flows of different ages
at Mono Lake, California have been differentiated and mapped by
their reflective spectral signatures including two specially filtered portions of the UV region. Multispectral scanner data of surface water bodies
near Jamestown, North Dakota were collected to determine the feasibility
of conducting more rapid and accurate surveys of the number and sizes
of suitable wildfowl breeding sites.

Studies concerned with spatial and temporal variations of ecosystem
components include the mapping of surface water temperature changes
on a river and the mapping of cooling rates of different Arctic terrain
features at Pt. Barrow, Alaska. Calibrated thermal IR detectors were
used to measure the rates of travel of cold water discharges from a
dam on the Clinch River in Tennessee, and to measure the rates of
surface water cooling. Figure 3 shows two views of a portion of the
Clinch River near a power plant 20 miles downstream from a dam.
In the top image, the warmest water entering the river from the plant
appears as a light grey tone; the coldest water inside a retaining wall
appears as a black tone; and the remainder of the river water which
is intermediate in temperature appears as a medium grey tone. The
bottom image was taken 23 hours after cold water was released at the
dam 20 miles upstream (Kauth 1968). A surface temperature map

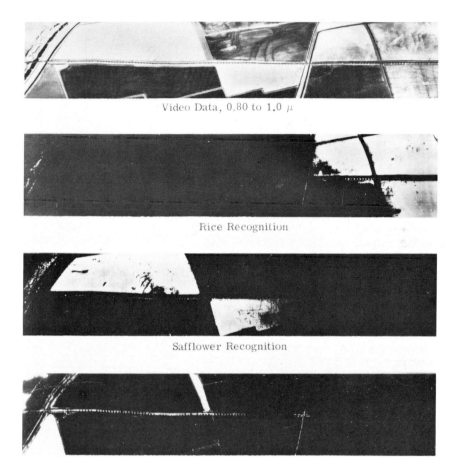

Video Data, 0.80 to 1.0 μ

Rice Recognition

Safflower Recognition

Bare Soil Recognition

Figure 2. Recognition Maps of Rice, Safflower, and Soil for Davis, California
Agricultural Area May 26, 1966 at 1400 hrs., 2000 ft.

of this area has been produced by enhancement techniques which will
be discussed in a later section.

Differences in the rates of cooling of selected Arctic terrain features is evident from examination of Figure 4. The imagery was
acquired during the transition period between summer and winter which
is characterized by unstable thermal conditions. The tundra surface
was frozen and snow covered at this time, and ice had formed on
the inland lakes. The warmest features in the scene appear as light
grey tones; the coldest features appear dark grey in tone. The light
grey tones of the inland lakes and the low center polygons indicate
the presence of water beneath the surface of the snow and ice which
serves as a heat source (Horvath 1968).

Sept. 3, 11:15 AM

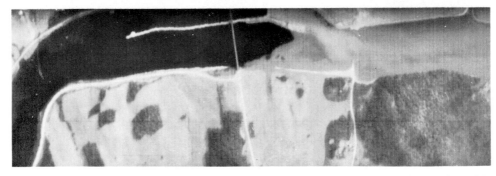

Sept. 4, 10:18 AM   8 to 14 μ

Figure 3. Monitoring of Water Temperature Changes on the Clinch River, Tennessee.

Examples of investigations concerning biotic or physical factors which control biological productivity include the mapping of moisture stressed apple trees and ponderosa pine trees. Figure 5 illustrates the comparative capability of the 0.32 - 0.38μ and the 2.0 - 2.6μ spectral bands for the detection of a moisture stressed condition of apple trees induced by rodent girdling at Ann Arbor, Michigan. In the 0.32 - 0.38μ image the damaged trees appear lighter in tone than the healthy trees, whereas no difference in tonal values of the damaged and healthy trees is noted in the 2.0 - 2.6μ image. Ground inspection of the orchard revealed that the damaged apple trees had smaller and fewer leaves and consequently more exposed branchwood. Laboratory spectrometer measurements showed that in the 0.32-0.38μ region the difference in mean reflectance values for abaxial surfaces of healthy and damaged leaf samples was 0.5%, and 1% for adaxial surfaces. The reflectance values of bark samples ranged from 3 to 7% higher than the reflectance values of the leaf samples in the 0.32-0.38μ region. The damaged apple trees were therefore differentiated from the healthy trees in the 0.32-0.38μ imagery by their highly reflective branchwood. The lack of image contrast between

8 – 13.5μ

Figure 4. Differential Cooling Rates of Terrain Features at Point Barrow, Alaska

0.32 - 0.38μm

2.0 - 2.6μm

Figure 5. Detection of Moisture Stressed Apple Trees

healthy and damaged trees in the 2.0-2.6μ region demonstrated that the reflectance values for both leaf and bark materials were nearly equal.

A typical procedure for the coordinated planning and research efforts of ecologists and remote sensing investigators involving nine separate tasks or steps is illustrated below. The procedure is described by the following example of an investigation underway at a forestry test site near Deadwood, South Dakota for the purpose of detecting the intensity and extent of beetle-infested ponderosa pine trees (Heller 1968). The investigation is incomplete, but the procedure used is typical.

Step 1: Definition of Objectives of Investigation

1. To determine if a temperature difference or non-visual reflectance difference between healthy and beetle-infested ponderosa pine trees could be detected by multispectral scanning techniques before any visual differences are observed.
2. To determine the accuracy of detection of beetle infested trees by comparison of scanner results with ground surveys.

Step 2: Physical Manifestations of Parameters of Interest

1. Temperature: Ground measurements of foliage temperatures indicated that the infested trees were 3° - 8° warmer than the healthy trees due to decreased rates of transpiration.
2. Color: Foliage of infested trees showed discoloration within one year after beetle attack.

Step 3: Sensor Characteristics Required to Solve Problem

1. Spectral regions: Calibrated thermal IR scanner data in the 4.5-5.5μ and 8.0-13.5μ spectral bands was required; and additional spectral coverage in the 0.4-2.6μ region.
2. Temperature resolution: 1°C temperature resolution was required in the thermal IR bands.

Step 4: Capability of Existing Sensors to Solve Problem

Existing sensors could provide calibrated thermal IR scanner data in the desired spectral bands with a temperature resolution less than 1°C.

Step 5: Sensor or Modification

Modification of existing sensors or design of new sensors are made in response to user needs for the solution of specific problems. This step was not considered necessary for the present investigation.

Step 6: Definition of Data Collection Program

Aerial-based operations

1. Flight schedule: May 29 and 30, 1968 at four time periods (early morning, and mid-morning, noon, and early afternoon) to obtain data under different levels of transpiration.
2. Weather conditions: bright sunshine with less than 30% cloud cover.
3. Position and lengths of flight paths: the flight paths had to cover ground instrumented sites and extend for two miles.
4. Number and order of data runs over each target area: six to eight data runs were requested for each target area.
5. Altitude of flights: 800 ft. above terrain was requested for all data flights.
6. Scanner configuration: 12 channel spectrometer (.40 - 1.0μ), a three element detector (1.0 - 1.4, 1.5 - 1.8, and 2.0 - 2.6μ), and a single element detector (8.0 - 13.5μ).
7. Calibration of thermal data: the thermal reference plates were set to register a 5 to 10°C temperature range without signal clipping; thermal ground resolution panels were overflown.
8. Calibration of reflective data: a "sun sensor" was used to record the intensity of total solar irradiance.

9. Radio communication: handie-talkie *Airnet* radios were used to coordinate timing and alignment of each run of the flight mission.

Ground-based operations

1. Temperature measurements: (a) five healthy and five insect-infested trees were instrumented for foliage absolute temperature; (b) apparent emitted temperature measurements were made of the thermal ground resolution panels; (c) radiometric surface temperatures of soil and other backgrounds of the trees were made from a stationary tower before, during, and after each flight.
2. Foliage moisture measurements: needle moisture tension measurements were made from samples of foliage collected from three trees of both healthy and dying classes.
3. Soil moisture measurements: soil pits were dug to determine water holding capacity and current field capacity.
4. Micrometeorological measurements: weather data from anemometers, pyreheliometers, hygrothermographs, and thermometers were collected in order to define the radiation environment.
5. Insect Infestation Measurements: type, extent, and intensity of insect infestation data were collected on a sample basis.
6. Foliage color measurements: Munsell color notations were made for all infested trees on five target sites.
7. Flight path marking devices: ground panels and lights were established for airborne orientation.

Step 7: Definition of the Data Reduction Program

The data reduction program has not been completely defined for this investigation. Typical considerations for this type of investigation include the following:

1. Evaluation of quality of the taped scanner data.
2. Evaluation of the "ground truth" data.
3. Definition of the purpose of the automatic signal processing and analysis (e.g., the differentiation and mapping of insect infested pine trees).
4. Selection of scanner data for automatic signal processing and analysis (i.e., selection of data runs and times, selection of spectral channels).
5. Analysis of scanner data to determine the most promising processing technique(s).
6. Selection of the most promising automatic signal processing technique(s) and form(s) of display of processed data.
7. Processing of selected scanner data to discriminate objects of interest.

Step 8: Processor Design or Modification

This step would be taken if the existing processing facilities are found to be inadequate for the analysis of a particular user problem either before or after Step 7 had been taken, or after Step 9 had been taken.

Step 9: Analysis and Evaluation of Sensor and Processor Outputs

Again for this investigation the data analysis tasks have not been completely defined, but will likely include the following:

1. Correlation of physical characteristics of objects sensed as indicated by ground truth data with observed differences in reflectance or temperature recorded in the scanner data.
2. Determination of the statistical reliability of the results of processed scanner data, and the need for modification of existing processing equipment or design of new processors.
3. Determination of the influence of scanner system variables (e.g., electronic noise, accuracy of calibration references) upon the quality of the sensor and processor outputs, and the need for scanner system modification or design.
4. Determination of the influence of illumination conditions (e.g., sun angle, cloud cover) on the quality of the sensor outputs.
5. Determination of modifications needed for future aerial-based and ground-based data collection programs to solve this particular ecological problem.
6. Reporting of results of the analysis of sensor outputs and processed data.

An important feature of the nine-step procedure described above is its provision for re-definition of tasks and re-evaluation of processing and analysis techniques. We believe that such feedback originating from both ecologists and remote sensor investigations is necessary for the improvement of remote sensing techniques and the more effective solution of ecological problems.

EXAMPLES OF RESULTS USING MULTISPECTRAL PROCESSING TECHNIQUES

Electronic processors have been used at Michigan to perform recognition and classification of signals from several areas for which multichannel scanner data have been collected. Since the data are produced and recorded in analog form, the first processing of the data was performed by using a simple two-channel analog technique. Based on experience with this technique and on theoretical analyses, more complex decision rules were implemented, rules which make use of the likelihood ratio of statistical decision theory. A special-purpose analog computer which makes decisions at a real-time rate was designed and fabricated and is now in operation. Equivalent decision operations have also been performed at a much slower rate on a digital computer from digitized samples of multichannel data, and research is being directed toward hybrid computers which will combine the speed of analog devices with the flexibility and control of digital computers.

Selected examples of processed multispectral data are discussed to illustrate the capability of multispectral systems to recognize crop species, detect differences in the maturity and vigor of crops, detect differences in water depths in shallow waters, and detect temperature differences in river water near a power plant. A variety of ways of presenting the data was employed.

The first example illustrates the recognition of crop species with the simple two-channel processor. Shown in Figure 2 (Thomson 1967) with a conventional image for the spectral band 0.4 to 1.0μ are three black and white recognition maps for an agricultural area in California. The recognition map for rice (produced using data in the channels 0.7-0.80μ) exhibits a good percentage of detection in the two rice fields, except for areas in the lower left-hand corner of each which may have had different soil moisture levels or different amounts of plant cover than the rest of the fields. The recognition map for safflowers (produced using data in the channels 0.48-0.50μ and 0.62-0.66μ) also exhibits a high detection percentage and correctly rejects the irrigation canals or dikes which have spectral characteristics that differ from those of the crop; portions of the lower field appear different from the remainder and were rejected for unknown reasons. The bare-soil recognition map (produced using data in the channels 0.48-0.50μ and 0.58-0.62μ) shows detection of gravel road shoulders along the highway and dirt roads between some of the fields in addition to the field of bare soil. It has been found very helpful to produce color-coded recognition maps which simultaneously present recognition data for several different crops. For example, the areas classified in Figure 2 as bare-soil could be brown, those for rice green, and those for safflowers red on a single color image. The remaining four fields were relatively immature rice which also have been differentiated by signal processing techniques.

It is also possible to produce a symbol on a digital computer printout to signify detection of a crop species (Hasell, et al. 1968). Figure 6a is such an output decision map for wheat. As can be seen from the ground-truth tables and the ground-truth map of Figure 6b, all wheat fields were correctly identified. The skewness of the computer printout is due to crabbing of the aircraft during overflight, and the horizontal compression of the fields is due to the sampling interval used in producing the digital data. It is possible to correct electronically for such distortion. The panchromatic photograph which serves as the background for the ground-truth map (Figure 6b) exhibits very little difference between fields of wheat and oats, yet by using data in several bands simultaneously these fields are easily distinguished from each other. The only potential false detection, field *SmG17*, was identified only as small grains and probably also contained wheat. A maximum likelihood decision rule with twelve channels of data in the region 0.4 to 1.0μ was used in the processing.

This multispectral approach may also aid in the mapping of underwater features and may enhance the study of aquatic ecosystems (Polcyn 1968). Figure 7 shows selected imagery taken September 7, 1967 from 10,000 feet of an area near Elliot Key on the southeast coast of Florida. Light penetration into the water was good in the 0.52

to 0.58μ region. As expected, the best water/land boundary was mapped in the 0.8 to 1.0μ region where water strongly absorbs the incoming light. In the 0.62 to 0.66μ band, only the shallowest parts of the coral reefs are seen because of relatively stronger absorption of water with respect to the 0.55 to 0.58μ region. In the region 0.40 to 0.44μ an image of lower contrast was produced because of the greater scattering by the water in this region. The capability for identifying underwater objects will be impaired because of the wavelength filter effect of the water and light attenuation caused by suspended materials. Nevertheless, the roles that the multispectral sensors can play are being developed for pollution studies, for depth measurements in shallow water and for measurements of chlorophyll concentrations in water.

The capability of new techniques for processing single-channel data can be illustrated by a thermal image (Figure 8a) of a power plant on the Clinch River in Tennessee (Kauth 1968). Because of the calibration of the sensing system, apparent temperature levels or differences can be measured and displayed by special procedures. Figure 8b is the analog video map quantitized into 16 grey levels. Enhancement of each of the 16 levels by superimposing a spike at the decision point between levels resulted in a temperature contouring for which we know the temperature interval represented by each grey tone and the spatial extent of surface in that temperature interval (Figure 8c). Similar schemes can be done with a digital computer with typed symbols on a printout instead of a grey tone produced on a film.

SUMMARY

The potential of multispectral remote sensing for ecological research lies in its ability to provide information about an object without necessarily resolving its spatial properties. This is particularly true for the recognition of a plant species by its multispectral signature, or the detection of physiological changes of plants by observing differences in their multispectral responses.

In order to accomplish this form of data collection reliably, new calibrated, synchronized, multichannel sensors are being developed. Results thus far show that whenever spectral differences occur and can be related to a parameter of interest about an object, multispectral sensing can aid in its detection, the mapping of its distribution, or the analysis of its energy budget characteristics.

A wide variety of special signal processing techniques are available to aid in the recognition and mapping of objects of interest and the measurement of energy budget parameters. Much work remains to understand the influence of instrumental and environmental limiting factors upon the statistical reliability of the automatic signal processing techniques. What is needed now is a close cooperation between the instrument and processor developers and the ecologists to determine by experiment, the degree to which multispectral sensing can contribute to a better understanding of the biological world.

RECOGNITION MAP FOR WHEAT
Maximum Likelihood Processing of Digitized 12-Channel Scanner Data
(Skewness due to crabbing of aircraft and sampling distortion)

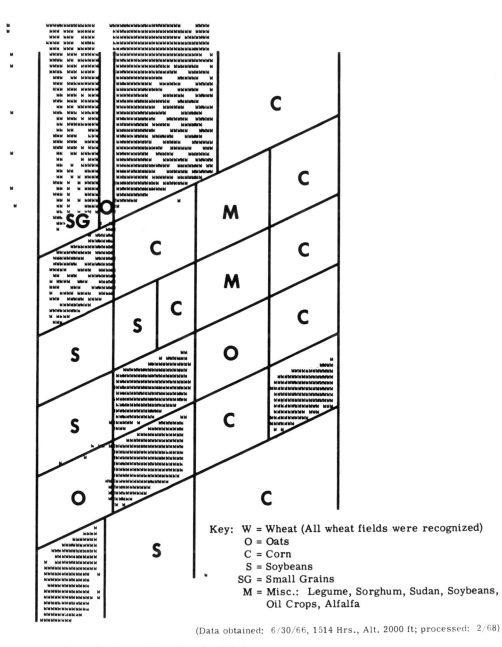

Key: W = Wheat (All wheat fields were recognized)
      O = Oats
      C = Corn
      S = Soybeans
      SG = Small Grains
      M = Misc.: Legume, Sorghum, Sudan, Soybeans,
                  Oil Crops, Alfalfa

(Data obtained: 6/30/66, 1514 Hrs., Alt. 2000 ft; processed: 2/68)

Figure 6a. Recognition Map for Wheat.

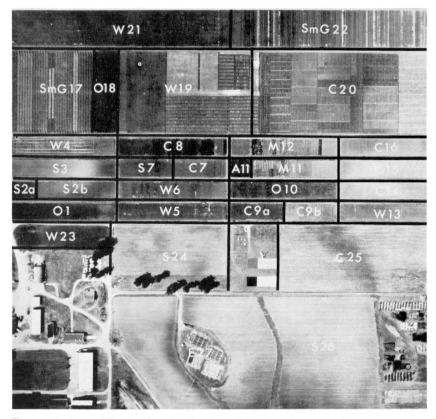

Figure 6b. Ground-Truth Map for Portion of Purdue Agronomy Farm.

We wish to acknowledge the support given by various agencies for the development of the multispectral instrumentation and the collection of data presented in this paper. The agencies concerned are the Department of the Interior, Department of Agriculture, The Arctic Institute of North America, Department of the Army, National Aeronautics and Space Administration, the Tennessee Valley Authority, and the Naval Oceanographic Office.

REFERENCES

Hasell, P. G. 1968. Investigations of spectrum matching techniques for remote sensing in agriculture, Report No. 8725-13-P, Willow Run Laboratories of the Institute of Science and Technology, Univ. Michigan, Ann Arbor.

Heller, R. C. 1968. Previsual detection of ponderosa pine trees dying from bark beetle attack. Proc. Fifth Symp., Remote Sensing Environment, Univ. Michigan, Ann Arbor, pp. 387-434.

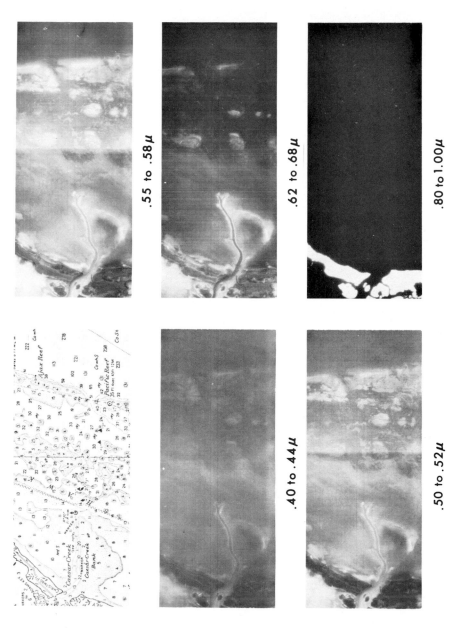

.55 to .58μ

.62 to .68μ

.80 to 1.00μ

.40 to .44μ

.50 to .52μ

Figure 7. Multispectral Comparison of Light Penetration of Water

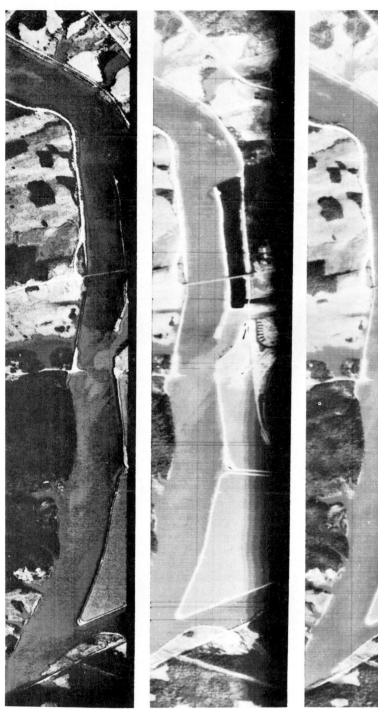

Figure 8. Map of Temperature Contours on the Clinch River, Tennessee.

Horvath, R. 1968. Multispectral survey of arctic regions, (Final Report), Report No. 1248-1-L, Institute of Science and Technology, Univ. Michigan, Ann Arbor.

Kauth, R. J. 1968. Preliminary analysis of TVA multispectral data, Report No. 1195-2-L, Institute of Science and Technology, Univ. Michigan, Ann Arbor.

Lowe, D. S., J. G. Braithwaite, and V. L. Larrowe. 1966. An investigative Study of a spectrum-matching imaging system, (Final Report), Report No. 8201-1-F, Institute of Science and Technology, Univ. Michigan, Ann Arbor.

Malila, W. 1968. Multispectral techniques for contrast enhancement and discrimination, Photogr. Engr. 34:566-575.

Polcyn, F. C. 1967. Investigations of spectrum-matching sensing in agriculture, (Semi-Annual Report), Report No. 6590-7-P, Institute of Science and Technology, Univ. Michigan, Ann Arbor.

Polcyn, F. C. 1969. Remote sensing techniques for the location and measurement of shallow-water features, Report No. 8973-10-F, Institute of Science and Technology, Univ. Michigan, Ann Arbor.

Steiner, D. and H. Hafner. 1965. Tone distortion for automated interpretation, Photogr. Engr. 31:269-280.

Thomson, F. J. 1967. Multispectral discrimination of small targets, Report No. 6400-135-T, ECOM-00013-135, Institute of Science and Technology, Univ. Michigan, Ann Arbor.

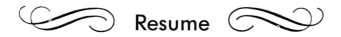 # Resume

PHILIP L. JOHNSON

HISTORY suggests that science has progressed according to its ability to measure smaller or larger phenomena. Remote sensing is a yardstick for environments, populations and ecosystems. Although photography is not new to science nor usually used quantitatively, recent technological advances have created new tools for ecologists and resources managers. Perhaps aerial photography and remote sensor imagery have the potential in ecology that the spectrophotometer has demonstrated in physiology. In fact aerial photography may be thought of at the ecosystem level as the converse of the electron microscope in molecular biology; each depends upon the spectral reflectance, absorption, emission and transmission characteristics of the respective samples. The exciting prospect is that remote sensing will be a logical bridge between intensive ecological research and its application to better planning and management of landscapes.

Photography from the ground or from aerial platforms is a familiar and conventional form of remote sensing that has been extensively used by foresters, geologists and geographers since the 1930's to inventory timber stands, to map geological structures, and to document landuse patterns. Technological advances in physics and electronics stimulated by military and space research have created a diversity of sensors that capture energy from various portions of the spectrum. Appropriate systems analyze and display the data in pictorial format or on tape for input to a computer. In spite of recent successes in automatic target recognition, the meaning and validity of remote sensing data require familiarity with the ecosystem recorded as well as with the basic-matter-energy relationships responsible for the images to be analyzed.

There are three basic types of systems available for remote sensing from airborne or satellite platforms. All three are usually processed to present two dimensional or pictorial displays: photography in the near visible spectrum 380 to 1000 mμ; optical-mechanical scanners from the ultraviolet through infrared wavelengths, 0.3 to 40μ; and passive microwave and radar for selected bands from 1 mm up to 1 meter. Radar is an active system in the microwave frequencies in which the appropriate energy is generated in the aircraft and directed toward the ground.

The radar return or reflected signal is captured by an antenna specific to that wavelength. The intensity of the energy returned is primarily a function of terrain aspect relative to beam direction and secondarily related to the dielectric properties of the reflecting material (Moore and Simonett 1967). The advantages of radar imaging are its independence of weather and diurnal conditions.

Frequently information from several spectral bands surpasses the sum of each band considered separately. This has led to the development of "multispectral" sensing in which several or many spectral energies are recorded simultaneously. The data gathered may be telemetered or returned as film or magnetic tape. Such information is a function of the environment imaged or plant and animal surfaces, or a complex interaction of both. The biological implications of the signals recorded must be analyzed and interpreted, and there often is little prior experience to guide the interpreter.

The bottleneck in extracting information from aerial photographs or scanner images has been the interpreter. A major advance was recently achieved with the fabrication of a single aperature, optional-mechanical-scanner with 12 channels. This was accomplished by combining a spectrophotometer with a conventional scanner (Holter and Wolfe 1960) such that energy entering an entrance slit is split into 12 channels from 400 to 1000 mμ. Each spectral band is recorded separately on magnetic tape and each channel is calibrated such that actual energy units received from the scene can be generated for each tone. Six additional channels, one in the ultraviolet and five in the infrared have been added with appropriate detectors.

Since each channel is recorded simultaneously, all images are in perfect register. The data for each spectral band or for selected bands can form input to a computer for correlation with "ground truth" or can be displayed as separate pictorial images for study. For the first time the ecologist can obtain multispectral data shortly after an overflight and can utilize it in quantitative form without the laborious task of photointerpretation. Electronic processers and analog computers can now perform recognition and classification assignments on multispectral data. Weak signals can be enhanced (Lowe and Braithwaite 1966) ; strong or irrelevant data can be masked.

Ecological research and planning are among the few alternatives for the survival of man's economic and cultural society that are increasingly stressed by the numbers of men and power of his technology. Ecology as a scientific discipline has an opportunity, indeed a mandate, to bridge the gap between academia, technology and the very real and extensive environmental problems before modern and future society. Toward this goal four kinds of ecological inquiry are susceptible to remote sensing techniques:

(1) Inventory and Mapping of Resources
(2) Quantizing the Environment
(3) Describing the Flow of Matter and Energy in the Ecosystem

(4) Evaluating Change and Alternative Solutions for Management of Ecosystems

Inventory and Mapping

Solutions of many resource problems depend on adequate assessment of physical and biological characteristics integrated over areas of a few square meters to thousands of kilometers. Maps of these characteristics are logical forms of communication. This approach is well illustrated by advances in geological exploration including magnetometer surveys and in forest inventories (Colwell 1961, 1968). Important biological properties of ecosystems potentially measureable by remote sensing techniques, singly or in multispectral combinations, include: leaf area index, stem volume, crop acreage and yield, species and structural diversity (Olson 1964, Miller 1960, Wickens, 1966), weight and chlorophyll content of vegetation; certain disease and insect infestations (Norman and Fritz 1965); kind, density and biomass of larger animal populations; thermal and chemical pollution of aquatic systems (Schneider 1968, Strandburg 1966); heat, water vapor and carbon dioxide fluxes of the earth's surfaces; evapotranspiration; and water content of soils and vegetation; fire (Bjornsen 1968); and depth and density of snow.

Quantitizing the Environment

In addition to what it is and where it is, most resource problems require quantitative data. In some cases even estimates of magnitude applied to large areas are sufficient for meaningful planning (Haefner 1967). In more sophisticated treatments, remote sensing, particularly the narrow bandpass instruments, can generate accurate numbers. The recent development of a laser system as an airborne profilometer for studies of microtopography appears capable of portraying changes in elevation on the order of a few centimeters from airplane altitudes (Rempel and Parker 1965). Radio ice sounding techniques permit measurements of the ice/bedrock interface through ice thicknesses of 2500 meters (Rinker et al. 1966). The recent development of sensitive heat and magnetic sensing tools is a product of military necessity which is rapidly finding applications in the civilian economy. One of the problems in this context has been the rapid increase in systems development and the resulting lag in man's knowledge on how such systems can be effectively utilized and with what degree of reliability. Consider the number of thermometers required to delineate cold air drainage in a valley versus an isothermal contour map by remote sensing.

Flow of Matter and Energy

Failure to understand the processes active in nature is frequently the cause of faulty resource planning. To appreciate function as opposed to structure in ecosystems requires the study of processes, interactions and transfer rates between different organisms as well as between organisms and environment. Frequently, measures of metabolic activity are desired. These are the most difficult answers to obtain from remote sensors as well as on the ground. Nevertheless, detection by airborne sensors of metabolic products or physical changes caused by active biological processes are important clues for inferences and conclusions about the direction and quantity of matter and energy flow in ecosystems (Barringer *et al.* 1968, Lohman and Robinove 1964).

Evaluating Change and Alternative Solutions

Alteration of natural ecosystems is, of course, manifest in all resources problems. Without change — that is depletion, erosion, pollution, accrual, or epidemic — the problem is seldom recognized. This, perhaps, is the easiest type of information to procure by repetitive aerial surveillance and has been exploited with photography in the visual wave lengths. Until repetitive aerial reconnaissance is widely practiced (and financed!), long term and widespread change, man caused or natural, will be difficult to assess. Once trends of change or the consequencese of our technology are evaluated, alternatives can often be developed from the same data. This is perhaps the most promising application for photography from spacecraft (Brock *et al.* 1965, Lowman 1966).

A major contribution of the technology developed to explore space will be to provide information to the natural sciences from earth orbiting spacecraft (Badgley and Vest 1966). Already satellites are telemetering valuable synoptic coverage of weather patterns. Once a spacecraft is in orbit the cost of repetitive photography is minimal. Some hand-held color transparencies obtained on the Gemini missions have shown the value and quality of present space photography. For example, the swirl in the Sargasso Sea had never been observed prior to Gimini photography. As systems are improved and designed specifically for space platforms, improvements in quality can be expected. Perhaps the most valuable and immediate return from satellite platforms will be synoptic photographic coverage of broad areas and changing conditions.

The solution of man's confrontation with nature involves many decisions outside the province of an ecologist. To the extent that many of these decisions are rooted in biological concepts it is worth contrasting multiple-use management with ecosystem zoning. It is apparent that we can no longer afford as many multiple uses of some landscapes. Perhaps assigning priorities and zones of use founded on understanding the ecosystem aided by remote sensing tools is a plausible alternate to irreversible multiple abuse. Realization of such apparent benefits through remote sensing will require far greater coordination, planning and operational capability than now exist or are proposed. Major technical problems have been solved; new approaches to ecological problems are posed; a valuable tool is emerging.

REFERENCES

Badgley, P. C. and W. L. Vest 1966. Orbital remote sensing and natural resources. Photogr. Engr. 32:780-790.

Barringer, A. R., B. C. Newbury and A. J. Moffat. 1968. Surveillance of air pollution from airborne and space platforms. Proc. 5th Symp., Remote Sensing of Environment, 123-156. Univ. Michigan, Ann Arbor.

Bjornsen, R. L. 1968. Infrared mapping of large fires. Proc. 5th Symp., Remote Sensing of Environment, 459-464. Univ. Michigan, Ann Arbor.

Brock, G. C., D. I. Harvey, R. J. Kohler, and M. P. Myskowski. 1965. Photographic considerations for aerospace. Itek Corp., Lexington, Mass., 118 pp.

Colwell, R. N. 1961. Some practical applications of multiband spectral reconnaissance. American Scientist 29:3-36.

Colwell, R. N. 1968. Remote sensing of natural resources. Scientific American 218:545-569.

Haefner, H. 1967. Airphoto interpretation of rural land use in Western Europe. Photogrammetria 22:143-152.

Holter, M., and W. Wolfe. 1960. Optical-mechanical scanning techniques. Univ. Michigan, Infrared Lab. 2900-154-R.

Lohman, S. W. and C. J. Robinove. 1964. Photographic description and appraisal of water resources. Photogrammetria. 19:21-41.

Lowe, D. S. and J. G. N. Braithwaite. 1966. A spectrum matching technique for enhancing image contrast. Applied Optics 5:893-898.

Lowman, P. D., Jr. 1966. The earth and orbit. Natl. Georgr. Mag. 645-671.

Miller, R. G. 1960. The interpretation of tropical vegetation and crops on aerial photographs. Photogrammetria 16:230-240.

Moore, R. K., and D. S. Simonett. 1967. Radar remote sensing in biology. BioScience 17:384-390.

Norman, G. G. and N. L. Fritz. 1965. Infrared photography as an indicator of disease and decline in citrus trees. Proc. Fla. State Horticult. Soc. 78:59-63.

Olson, D. P. 1964. The use of aerial photographs in studies of marsh vegetation. Maine Agr. Expt. Sta. Bull. 13, 62 pp.

Remple, R. C. and A. K. Parker. 1965. An information note on an airborne laser terrain profiler and micro-relief studies. Proc. 3rd Symp., Remote Sensing Environment, 321-337. Univ. Michigan, Ann Arbor.

Rinker, J. N., S. Evans and G. de Q. Robin. 1966. Radio ice-sounding techniques. Proc. 4th Symp., Remote Sensing of Environment, 793-800. Univ. Michigan, Ann Arbor.

Schneider, W. J. 1968. Color photographs for water resources studies. Photogr. Engr. 34:157-162.

Strandburg, C. H. 1966. Water quality analysis. Photogr. Engr. 32:234-248.

Wickens, G. E. 1966. The practical application of aerial photography for ecological surveys in the savannah regions of Africa. Photogrammetria 21:33-41.

Selected Bibliography

REMOTE sensing is a truly interdisciplinary tool. As such it attracts increasing interest from investigators whose training and experience may not encompass the fundamental technology or specific instrumentation appropriate to remote sensing. Therefore, a bibliography was selected to acquaint the interested reader with the varied literature in this expanding field with reference to application in ecology and natural resources. Emphasis is placed on references published since 1950, although earlier citations are included for historical perspective or for their treatment of pertinent matter-energy relationships.

Biotelemetry literature was reviewed in the February issue of Bio-Science and most recently (May 1968) by J. I. Schladweiler and I. J. Ball as Technical Report 15 of the Bell Museum of Natural History, University of Minnesota, Minneapolis, Minn.

Four concurrent publications compliment this symposium. The September 1968 issues of *Scientific American* and *Applied Optics* are respectively devoted to articles about light and about infrared remote sensing applications in nondestructive testing and the medical sciences. The proceedings of the Toulouse Conference (UNESCO. 1968. *Aerial Surveys and Integrated Studies*. Paris, 575 p.) presents 60 papers and six panel discussions from a symposium convened at Toulouse, France in September 1964 for an appraisal of the use of aerial photography for inventory of natural resources. The reader will find European references more readily in this volume. A *Manual of Aerial Color Photography* has recently been published (1968) by the American Society of Photogrammetry, Washington, D. C. Examples of color aerial photography are analyzed with respect to a number of fields of applied ecology.

The reader will appreciate that a considerable amount of remote sensing research has been developed by contracts from the Department of Defense. Many of the reports of this research, both in and out of government laboratories, remain classified. No classified literature is included in this bibliography. The Editor gratefully acknowledges the contributions to this bibliography by Drs. L. D. Miller, C. E. Olson and R. L. Marchinton.

Ahmed, N and W. W. Koepsel. 1960. An approach to correlate pulsed radar and photographic data. NASA CR-50536, 71 p.

Aldrich, R. C., W. F. Bailey, and R. C. Heller. 1959. Large scale 70 mm color photography techniques and equipment and their application to a forest sampling problem. Photogr. Engr. 25:747-754.

Aldrich, R. C. 1966. Forestry applications of 70 mm color. Photogr. Engr. 32:802-810.

Aldrich, R. C. 1968. Remote sensing and the forest survey: present applications, research and a look at the future. Proc. 5th Symp., Remote Sensing of Environment, 357-372. Univ. Michigan, Ann Arbor.

Alexander, R. H., L. W. Bowden, D. F. Marble, and E. G. Moore. 1968. Remote sensing of urban environments. Proc. 5th Symp., Remote Sensing of Environment, 889-912. Univ. Michigan, Ann Arbor.

Allen, W. A. and A. J. Richardson. 1968. Interaction of light with a plant canopy. Proc. 5th Symp., Remote Sensing of Environment, 219-232. Univ. Michigan, Ann Arbor.

Allison, G. W. and R. E. Breadon. 1960. Timber volume estimates from aerial photographs. British Columbia Forest Service, Forest Survey Note 5.

American Society of Photogrammetry. 1952. Manual of Photogrammetry Ed. 2. Banta Press, Menasha, Wis. 876 pp.

American Society of Phtotogrammetry. 1960. Manual of Photographic Interpretation. Banta Press, Menasha, Wis. 868 pp.

Antonini, G. A. 1965. Infrared image applications in studies of the marine environment. Proc. 3rd Symp., Remote Sensing of Environment, 781-788. Univ. Michigan, Ann Arbor.

Atkinson, J. H., Jr. 1963. Atmospheric limitations on ground resolution from space photography. 7th SPIE Technical Symposium, New York, N. Y. S. P. I. E. Journal, Vol. 1.

Atta, G. F. Van. 1936. Filters for the separation of living and dead leaves in monochromatic photographs with a method for determination of photographic filter factors. Biological Photographic Assoc. 4:177-191.

Austin, R. M. 1958. Aerial war against cereal crop diseases. Industrial Photography 7:38-40.

Avery, T. E. 1958. Helicopter stereo-photography of forest plots. Photogr. Engr. 24:617-625.

Avery, G. 1960. Identifying southern forest types on aerial photographs. U. S. Forest Service, Southeastern Forest Experiment Station. Paper No. 112.

Avery, T. E. 1962. Interpretation of aerial photographs. Burgess Publishing Company, p. 192.

Avery, T. E. and M. P. Meyer. 1959. Volume tables for aerial timber estimating in northern Minnesota. Lakes States Forest Expt. Sta. Paper 78.

Avery, T. 1966. Forester's guide to aerial photo-interpretation. U. S. Department of Agriculture, Department Handbook, 40 p.

Avery, T. E. and H. Meyer. 1962. Contracting for forest aerial photography in the United States. U. S. Department of Agriculture, Forest Service, Lake States For. Expt. Station, Sta. Paper 96.

Avery, T. E. and Richter. 1965. An airphoto index to physical and cultural features in eastern United States. Photogr. Engr. 31:896-914.

Axelson, Hans. 1056. Effect of photo scale on the use of aerial photographs in Swedish forestry. Norrlauds Skogsvardsforbunds Tidskriff, 252-292.

Babel, A. 1935. Infrarot-photographic in Pflanzenschutz. Angew. Botan. 17:43-53.

Backstrom, H. and E. Welander. 1953. En undersokning av Remissions-formagan has Blad och barr av Olika Tradslag (An investigation of the spectral remission power of leaves and needles of different tree species). Sartryck ur Norrlauds Skogsvardsforbunds Tidskrift 1:141-169.

Badgley, P. C. 1966. Current status of NASA's natural resources program. Proc. 4th Symp., Remote Sensing of Environment, 547-570. University of Michigan, Ann Arbor.

Badgley, P. C. and W. L. Vest. 1966. Orbital remote sensing and natural resources. Photogr. Engr. 32:780-790.

Badgley, P. C., A. P. Colvocoresses and C. D. Centers. 1966. NASA earth-sensing space flight experiments. Photogr. Engr. 34:162-167.

Barath, F. T. 1965. Microwave radiometry and applications to oceanography. Woods Hole Oceanographic Inst. Oceanography from Space, Ref. 65-10.

Barringer, A. R. 1963. The use of audio and radio frequency pulses for terrain sensing. Proc. 2nd Symp., Remote Sensing of Environment, 201-214. Univ. Michigan, Ann Arbor.

Barringer, A. R. 1966. The use of multi-parameter remote sensors an important new tool for mineral and water resource evaluation. Proc. 4th Symp., Remote Sensing of Environment, 313-325. Univ. Michigan, Ann Arbor.

Barringer, A. R. and J. P. Schock. 1966. Progress in the remote sensing of vapours for air pollution, geologic and oceanographic applications. Proc. 4th Symp., Remote Sensing of Environment, 779-792. Univ. Michigan, Ann Arbor.

Barringer, A. R., B. C. Newbury and A. J. Moffatt. 1968. Surveillance of air pollution from airborne and space platforms. Proc. 5th Symp., Remote Sensing of Environment, 123-156. Univ. Michigan, Ann Arbor.

Bawden, F. C. 1933. Infrared photography and plant virus diseases. Nature 132:168.

Becking, R. W. 1959. Forestry applications of aerial color photography Photogr. Engr. 25:559-565.

Belecher, J. 1948. Determination of soil conditions from aerial photographs. Photogr. Engr. 14:482-488.

Berstein, D. A. 1962. Guide to two-story forest type mapping in the Douglas-fir Sub-region. U. S. Forest Service, Portland, Oregon.

Belov, S. V. and E. S. Anteybasher. 1957. A Study of reflecting properties of arboreal species. Botaniches Zhurnal 42:517-534.

Billings, W. D., R. J. Morris, 1951. Reflection of visible and infrared radiation from leaves of different ecological groups. Amer. J. Bot. 38:327-331.

Birkebak, R. and Birkebak. 1964. Solar radiation characteristics of tree leaves. Ecology 45:646-649.

Bjornsen, R. L. 1968. Infrared mapping of large fires. Proc. 5th Symp., Remote Sensing of Environment, 459-464. Univ. Michigan, Ann Arbor.

Boesch, H. and D. Steiner. 1959. Interpretation of land utilization from aerial photographs. Final Tech. Rept. Geographisches Institute der Universitat Zurich, Zurich, 69 p.

Brenchley, G. H. and C. V. Dadd. 1962. Potato blight recording by aerial photography. N.A.A.S. Quarterly Rev., London 57:21-25.

Brock, G. C. 1952. Physical Aspects of Air Photography. Longmans, Green and Co., London 280 pp.

Brock, G. C., D. I. Harvey, R. J. Kohler, and M. P. Myskowski. 1965. Photographic considerations for aerospace. Itek Corp., Lexington, Mass. 118 p.

Brunnschweiler, D. H. 1957. Seasonal changes of the agricultural pattern: a study in comparative airphoto interpretation. Photogr. Engr. 23:131-139.

Buckhorn, W. J. and P. G. Lauterbach. 1957. Timing of aerial surveys for the balsam woolly aphid. Pacific Forest & Range Expt. Sta. Res. Note 142, 2 pp.

Buettner, K. J. K. and C. D. Kern. 1965. The determination of infrared emissivities of terrestrial surfaces. Jour. Geophys. Res. 70:1329-1337.

Bullock, P. 1956. Atmospheric haze in aerial photography. Photogr. Engr. 22:967-972.

Burks, G. F. and R. C. Wilson. 1939. A vegetative inventory from aerial photographs. Photogr. Engr. 5:30-42.

Bushnell, T. M. 1951. Use of aerial photographs for Indiana land studies. Photogr. Engr. 17:725-738.

Cain, S. A. 1966. Current and future needs for remote sensor data in ecology. Proc. 4th Symp., Remote Sensing of Environment, 3-6. Univ. Michigan, Ann Arbor.

Cameron, H. L. 1961. Interpretation of high-altitude small-scale photography. Canadian Surveyor 15:567-573.

Carman, P. D. 1951. Brightness of fine detail in air photography. J. Optical Soc. Amer. 41:305-310.

Carneggie, D. M. and D. T. Lauer. 1966. Uses of multiband remote sensing in forest and range inventory. Photogrammetria 21:115-141.

Clark, Walter. 1946. Photography by infrared. John Wiley and Sons, New York, N. Y., 472 pp.

Choate, G. A. 1961. Estimating douglas-fir site quality from aerial photographs. Pac. Northwest Forest Range Expt. Sta. Res. Pap. 45.

Ciesla, W. M., J. C. Bell, J. W. Curlin. 1967. Color photos and the Southern pine beetle. Photogr. Engr. 33:883-888.

Clarke, J. and R. B. Stone. 1965. Marine biology and remote sensing. Woods Hole Oceanographic Inst., Oceanography from Space, Ref. 65-10.

Clouston, G. 1950. The use of aerial photographs in range inventory work on the National Forests. Photogr. Engr. 16:329-331.

Colwell, R. N. 1950. New technique for interpreting aerial, color photography. J. Forestry 48:204-205.

Colwell, R. N. 1952. Determining relative illumination for overlapping photos for maximum stereoscopic effect. J. Forestry 50:369.

Colwell, R. N. 1954. A systematic analysis of some factors affecting photographic interpretation. Photogr. Engr. 20:433-454.

Colwell, R. N. 1960. Some uses of infrared aerial photography in the management of wildland areas. Photogr. Engr. 26:774-785.

Colwell, R. N. 1960. Some uses and limitations of aerial color photography in agriculture. Photogr. Engr. 26:220-222.

Colwell, R. N. 1961. Aerial photographs show range conditions. California Agriculture 15:12-13.

Colwell, R. N. 1961. Some practical applications of multiband spectral reconnaissance. American Scientist 29:3-36.

Colwell, R. N. 1963. Platforms for testing multi-sensor equipment. Proc. 2nd Symp., Remote Sensing of Environment, 7-49. Univ. Michigan, Ann Arbor.

Colwell, R. N. 1964. Aerial photography—a valuable sensor for the scientist. Amer. Scientist 52:17-49.

Colwell, R. N. 1965. Spectrometric considerations involved in making rural land use studies with aerial photography. Photogrammetria 20:15-33.

Colwell, R. N. 1965. Aids for the selection and training of photo interpretors. Photogr. Engr. 31:327-339.

Colwell, R. N. 1966. Uses and limitations of multispectral remote sensing. Proc. 4th Symp., Remote Sensing of Environment, 71-100. Univ. Michigan, Ann Arbor.

Colwell, Robert N. 1966. Aerial photography of the earth's surface, its procurement and use. Applied Optics 5:883-892.

Colwell, R. N. 1967. Remote sensing as a means of determining ecological conditions. BioSciences 17:444-451.

Colwell, R. N. 1968. Remote sensing of natural resources. Scientific American 218:545-69.

Colwell, R. N. 1968. Determining the usefulness of space photography for natural resource inventory. Proc. 5th Symp., Remote Sensing of Environment, 249-290. Univ. Michigan, Ann Arbor.

Colwell, R. N., W. Brewer, G. Landis, P. Langley, J. Morgan, J. Rinker, J. M. Robinson and A. L. Sorem. 1963. Basic matter and energy relationships involved in remote reconnaissance. Photogr. Engr. 29:761-799.

Colwell, R. N. and D. L. Olson. 1965. Thermal infrared imagery and its use in vegetation analysis by remote aerial reconnaissance. Proc. 3rd Symp., Remote Sensing of Environment, 607-617. Univ. Michigan, Ann Arbor.

Cook, J. C. 1963. Monocycle radar pulses as environmental probes. Proc. 2nd Symp., Remote Sensing of Environment, 223-231. Univ. Michigan, Ann Arbor.

Cooper, C. F. 1965. Potential applications of remote sensing to ecological research. Proc. 3rd Symp., Remote Sensing of Environment, 601-606. Univ. Michigan, Ann Arbor.

Cooper, C. F. and F. M. Smith. 1966. Color aerial photography: toy or tool. J. Forestry 64:373-378.

Costello, R. B. 1951. Possibilities of utilizing air photo interpretation in the Cornell economic land classification system. Land Economics 27:34-30.

Coulson, L. 1966. Effects of reflection properties of natural surfaces in aerial reconnaissance. Applied Optics 5:905-917.

Cross, B. 1962. Aerial photos: new weapon against pollution. Chemical Engineer, April 2, 1962, pp. 42-43.

Davis, K. and T. Neal. 1963. Descriptions and airphoto characteristics of desert landforms. Photogr. Engr. 29:621-631.

De Rosayro, R. A. 1959. The application of aerial photography to stock-mapping and inventories on an ecological basis in rain forests of Ceylon. Empire Forest Rev. 38, No. 96.

Dill, W. 1959. Use of the comparison method in agricultural airphoto interpretation. Photogr. Engr. 25:44-49.

Dill, W. 1963. Airphoto analysis in outdoor recreation: site inventory and planning. Photogr. Engr. 29:67-70.

Donner, W. 1963. Gas chromatography as a remote sensing device. Proc. 2nd Symp., Remote Sensing of Environment, 393-402. Univ. Michigan, Ann Arbor.

Doverspike, G. E. and R. C. Heller. 1962. Identification of tree species on large scale panchromatic and color photographs. Proc. of the International Symp. Photointerpretation, Delft, Netherlands.

Doverspike, G. E., F. M. Flynn and R. C. Heller. 1965. Microdensitometer applied to land use classification. Photogr. Engr. 31:294-306.

Drake, H. 1964. A spectral reflectance study using a wedge spectrograph. Photogr. Engr. 29:684-689.

Drury, H. W., I. C. T. Nisbet and R. E. Richardson. 1961. The migration of "angels" (birds and radar). Natural History 70:11-17.

Eastman Kodak. 1963. Infrared and Ultraviolet Photography. 7th ed. Rochester, N. Y.

Eastman Kodak. 1965. Kodak Wrattan Filters for Scientific & Technical Use. 22nd ed. Rochester, N. Y.

Eastman Kodak. 1967. Kodak Data for Aerial Photography. Kodak Publication No. M-29, Eastman Kodak Co., Rochester, New York, p. 83.

Eastwood, E. 1960. Radar ring "angels" and the roosting movement of starlings. Nature 186:112-114.

Eastwood, E. 1967. Radar Ornithology. Barnes and Noble, 278.

Eggert, J. 1935. Fortschritte und grenzleistunger in der Infrarot-photographie. Veroffentl. wiss. Zentral-Lab. phot. Abt. AGFA, 4:101-118.

Ellermeier, R. C. and D. S. Simonett. 1965. Imaging radars on space-craft as a tool for studying the earth. Univ. Kansas, CRES Rept. 61-66.

England, G and J. O. Morgan. 1965. Quantitative airborne infrared mapping. Proc. 3rd Symp., Remote Sensing of Environment, 681-690. Univ. Michigan, Ann Arbor.

Ewing, G. C. 1966. Current and future needs for remotely sensed ocean-ography data—a speculation. Proc. 4th Symp., Remote Sensing of Environment, 7-12. Univ. Michigan, Ann Arbor.

Falkner, H. 1960. Applying photogrammetry to range allotment planning. Photogr. Engr. 26:672-674.

Feder, A. M. 1960b. Interpreting natural terrain from radar displays. Photogr. Engr. 26:618-630.

Federer, C. A. and C. B. Tanner. 1966. Spectral distribution of light in the forest. Ecology 47:555-561.

Federer, C. A. and C. B. Tanner. 1966. Sensors for measuring light available for photosynthesis. Ecology 47:654-657.

Finley, V. P. 1960. Photo-interpretation of vegetation, literature survey and analysis. U. S. Army, Cold Regions Res. Engr. Lab. Tech. Report 69, 36 p.

Fischer, W. A. 1960. Reflectance measurements as a basis for film-filter selection for photographic differentiation of rock units. U.S.G.S. Professional Paper 400-B: 136-138.

Fischer, W. A., R. M. Moxham, F. Polcyn and G. H. Landis. 1964. Infrared surveys of Hawaiian Volcanoes. Science 146:733-742.

Francis, D. A. 1963. Aerial survey methods for forestry and forest industry pre-investment surveys in developing countries. Proc. Symp. Photo. Interpretation, Delft, 1962. p. 200-210.

Frey, H. T. 1967. Agricultural application of remote sensing—The potential from space platforms. U. S. Dept. of Agriculture, Economic Research Service, Agriculture Information Bulletin, No. 328, 28 pp.

Fricke, W. 1965. Falschfarben-Photographic fur die Luftbild-Interpretation. UMSCHAU in Wissenschaft und Technik 14:441.

Fritz, N. L. 1967. Optimum methods for using infrared sensitive color films. Photogr. Engr. 33:1128-1138.

Frost, R. E. 1950. Evaluation of soils and permafrost conditions in the Territory of Alaska by means of aerial photography. Purdue Engr. Expt. Sta., Vol. 1 and 2, 50, 112 pp.

Frost, R. E. 1953. Factors limiting the use of aerial photographs for analysis of soil and terrain. Photogr. Engr. 19:427-436.

Frost, R. E., P. L. Johnson, R. D. Leighty, V. H. Anderson, A. O. Poulin and J. N. Rinker. 1965. Selected airphoto patterns for mobility studies in Thailand. Vol I, Vegetation and surface composition. U. S. Army, Cold Regions Res. and Engr. Lab.

Fuchs, M. and C. B. Tanner. 1966. Infrared thermometry of vegetation. Agronomy Journal 58:597-601.

Fuchs, M., E. T. Kanemasu, J. P. Kerr and C. B. Tanner. 1967. Effect of viewing angle on canopy temperature measurements with infrared thermometers. Agronomy Journal 59:454-496.

Garver, R. and K. E. Moessner. 1949. Forest Service use of aerial photographs. Photogr. Engr. 15:504-517.

Gates, M. 1961. Winter thermal radiation studies in Yellowstone Park. Science 134:32-35.

Gates, D. M. 1965. Energy, plants and ecology. Ecology 46:1-13.

Gates, D. M. 1965. Characteristics of soil and vegetated surfaces to reflected and emitted radiation. Proc. 3rd. Symp., Remote Sensing of Environment, 573-600. Univ. Michigan, Ann Arbor.

Gates, D. M. 1967. Remote sensing for the biologist. BioScience 17:303-307.

Gates, D. M. and W. Tantraporn. 1952. The reflectivity of deciduous trees and herbaceous plants in the infrared to 25 microns. Science 115:613-616.

Gates, D. M., H. J. Keegan, J. C. Schlefer and V. R. Weidner. 1965. Spectral properties of plants. Applied Optics 4:11-20.

Gausman, H. W. and R. Gardenas. 1968. Effect of pubescence on reflectance of light. Proc. 5th Symp., Remote Sensing of Environment, 291-298. Univ. Michigan, Ann Arbor.

Gibson, H. L., W. R. Buckley, K. E. Whitmore. 1965. New vistas in infrared photography for biological surveys. J. Biol. Photo Assoc. 33:1-33.

Giever, P. M. 1966. Needs for remote sensing data in the field of air and water pollution control. Proc. 4th Symp., Remote Sensing of Environment, 21-24. Univ. Michigan, Ann Arbor.

Gimbarzersky, P. 1966. Land inventory interpretation. Photogr. Engr. 32:967-976.

Glover, K. M., K. R. Hardy, T. G. Konrad, W. N. Sullivan and A. S. Michaels. 1966. Radar observations of insects in free flight. Science 154:967-972.

Goodman, M. 1959. A technique for the identification of farm crops on aerial photographs. Photogr. Engr. 25:131-137.

Gordeer, R. V. 1961. Study of forest reserves by the method of axonemetric aerial photography. (Transl. from Russian). Office Tech. Serv. U. S. Dept. Commerce, Washington, D. C.

Gordon, J. I. and P. V. Church. 1966. Overcast sky luminances and directional luminous reflectances of objects and backgrounds under overcast skies. Applied Optics 5:919-924.

Guy, M. 1967. Quelques principes et quelques experiences sur la methodologie de la photo-interpretation. Proc. 2nd Symp. Internate. Photo-Interpretation Paris (1966), 21-41.

Haack, P. M. 1962. Evaluating color, infrared, and panchronatic aerial photos for the forest survey of interior Alaska. Photogr. Engr. 592-598.

Haefner, H. 1967. Airphoto interpretation of rural land use in Western Europe. Photogrammetria 22:143-152.

Haralick, R. M. 1968. Adaptive pattern recognition with a predictive criterion using agricultural radar images in western Kansas. Proc. 5th Symp., Remote Sensing of Environment, 343-356. Univ. Michigan, Ann Arbor.

Harris, D. E. and C. L. Woodbridge. 1964. Terrain mapping by use of infrared radiation. Photogr. Engr. 30:134-139.

Hegg, K. M. 1967. A photo identification guide for the land forest types of interior Alaska. Northern Forest Experiment Station, Res. Paper Nor-3.

Heinicke, R. 1963. The micro-climate of fruit trees. II. Foliage and light distribution patterns in apple trees. Proc. Amer. Soc. Horticultural Science, Vol. 83.

Heller, R. C., J. F. Coyne and J. L. Bean. 1955. Airplanes increase effectiveness of southern pine beetle surveys. J. Forestry 53:483-487.

Heller, R. C., R. C. Aldrich and W. F. Bailey. 1959. Evaluation of several camera systems for sampling forest insect damage at low altitude. Photogr. Engr. 25:137-144.

Heller, Robert C. and Donald Schmiege. 1962. Aerial survey techniques for the spruce budworm in the lake states. J. Forestry. 60:525-532.

Heller, R. C., G. E. Doverspike and R. C. Aldrich. 1964. Identification of tree species on large-scale panchromatic and color aerial photographs. U. S. Dept. Agr., Agricultural Handbook 261, 17 pp.

Heller, R. C., W. F. McCambridge, R. C. Aldrich, and F. P. Weber. 1966. The use of multispectral sensing techniques to detect ponderosa pine trees under stress from insect or pathogenic organisms. Pacific Southwest Forest and Range Experiment Station, Berkeley, California, 70 pp.

Heller, R. C. 1968. Previsual detection of ponderosa pine trees dying from bark beetle attack. Proc. 5th Symp., Remote Sensing of Environment, 387-434. Univ. Michigan, Ann Arbor.

Henriques, E. 1949. Practical application of photogrammetry in land classification as used by the Bureau of Land Management. Photogr. Engr. 15:540-548.

Herrington, R. B. and S. R. Tocher. 1967. Aerial photo techniques for a recreation inventory of mountain lakes and streams. U. S. Forest Service Res. Paper INT-37.

Highway Research Board. 1958. Air photo and soil mapping methods appraisal and application. Nat'l. Acad. Sci.-Nat'l. Res. Council Publ. 540, Washington, D. C.

Hills, G. A. 1950. The use of aerial photography in mapping soil site. For. Chronicle 26:4-37.

Hills, G. A. 1960. Regional site research. For. Chronicle. 36:401-423.

Hindley, E. and J. H. G. Smith. 1957. Spectrophotometric analysis of foliage of some British Columbia conifers. Photogr. Engr. 25:894-895.

Hirsch, S. N. 1963. Applications of remote sensing to forest fire detection and suppression. Proc. 2nd Symp., Remote Sensing of Environment, 295-308. Univ. Michigan, Ann Arbor.

Hirsch, S. N. 1965. Preliminary experimental results with infrared line scanners for forest fire surveillance. Proc. 3rd Symp., Remote Sensing Environment, 623-648. Univ. Michigan, Ann Arbor.

Hirsch, S. N. 1968. Project fire scan-summary of 5 years' progress in airborne infrared fire detection. Proc. 5th Symp., Remote Sensing Environment, 447-458. Univ. Michigan, Ann Arbor.

Hodgin, D. M. 1963. The characteristics of microwave radiometry in remote sensing of environment. Proc. 2nd Symp., Remote Sensing of Environment, 127-137. Univ. Michigan, Ann Arbor.

Hoffer, R. M., R. A. Holmes, and J. R. Shay. 1966. Vegetative, soil, and photographic factors affecting tone in agricultural remote multispectral sensing. Proc. 4th Symp., Remote Sensing of Environment, 115-134. Univ. Michigan, Ann Arbor.

Hoffer, R. M., C. J. Johannsen and M. F. Baumgardner. 1967. Agricultural applications of remote multispectral sensing. Proc. Indiana Acad. Sci. 76:387-395.

Holdridge, L. R., W. C. Grenke, W. H. Hatheway, T. Liang and S. A. Tosi. 1968. Forest environments in tropical life zones: a pilot study. Vol. I and II. Wilson, Nuttall, Raimond Engineers, Inc., 832 pp.

Holter, M. R. 1967. Infrared and multispectral sensing. BioScience 17:376-383.

Holter, M., and W. Wolfe. 1960. Optical-mechanical scanning techniques. Univ. Michigan, Infrared Lab. 2900-154-R.

Holter, M. and F. Polcyn. 1964. Comparative multispectral sensing. Univ. Michigan, Infrared Lab., 2900-484-S.

Holter, M. and R. Leqault. 1965. The motivation for multispectral sensing. Proc. 3rd Symp., Remote Sensing of Environment, 71-78. Univ. Michigan, Ann Arbor.

Huddleston, H. F. and E. H. Roberts. 1968. Use of remote sensing for livestock inventories. Proc. 5th Symp., Remote Sensing of Environment. 307-324. Univ. Michigan, Ann Arbor.

Hunkins, K. L., M. Ewing, B. C. Heeyen and R. I. Menzies. 1960. Biological and geological observations on the first photographs of the Arctic Ocean deep-sea floor. Limnol. Oceanogr. 5:154-161.

Hunt, G. R., J. W. Salisbury and J. W. Reed. 1967. Rapid remote sensing by spectrum matching techniques. 2. Application in the laboratory and in lunar observations. J. Geophys. Res. 72:705-719.

Ives, H. E. 1927. Airplane Photography. J. B. Lippincott, Philadelphia. 422 pp.

Ives, R. L. 1939. Infrared photography as an aid in ecological surveys. Ecology 20:433-439.

International Union of Forestry Research Organizations. 1963. Aerial photographs in forest inventories, applications and research studies, 1962. International Union Forest Research Organizations, Munich 96 p.

James, T. H. and J. F. Hamilton. 1965. The photographic process. International Scientific Technology: 38-44.

Jensen, and E. F. Peterson. 1948. Prospecting from the air. Scientific American. 24-

Jensen, H. A. and R. N. Colwell. 1949. Panchromatic versus infrared minus-blue aerial photography for forestry purposes in California. Photogr. Engr. 15:201-223.

Jensen, N. P. 196. Optical and Photographic Reconnaissance Systems. John Wiley and Sons, New York, N. Y.

Johnson, P. L. 1965. Radioactive contamination to vegetation. Photogr. Engr. 31:984-990.

Johnson, P. L. 1965. Investigation of sugar cane vigor with aerial photography in Puerto Rico. U. S. Army, Cold Regions Res. Engr. Lab., special Rept. 93:1-38.

Johnson, P. L. 1966. A consideration of methodology in photo interpretation. Proc. 4th Symp., Remote Sensing of Environment, 719-725. Univ. Michigan, Ann Arbor.

Johnson, P. L. 1968. Remote sensing as an ecological tool. UNESCO Symp. Ecology of Subarctic Regions, Helsinki (1966), (In Press).

Johnson, P. L. and T. C. Vogel. 1966. Vegetation of the Yukon Flats Region, Alaska. U. S. Army, Cold Regions Res. Engr. Lab, Research Rept. 209. 1-53.

Johnson, P. L. and D. M. Atwood. 1969. Aerial sensing and photographic study of the El Verde rain forest, Puerto Rico. In A Rain Forest, H. T. Odum (editor). U. S. Atomic Energy Comm. (In Press)

Keegan, J. and T. O'Neil. 1951. Spectrophotometric study of autumn leaves. Proc. Optical Soc. Am. 41:284.

Keegan, J., J. C. Schleter and W. A. Hall. Spectrophotometric and colorimetric change in the leaf of a white oak tree under conditions of natural drying and excessive moisture. Natl. Bureau of Standards, Report 4322, U. S. Dept. Commerce.

Keegan, J., J. C. Schleter, W. A. Hall and G. M. Spectrophotometric and colorimetric study of foliage stored in covered metal containers. Natl. Bureau of Standards, Report 4370, U. S. Dept. of Commerce.

Keegan, H. J. 1957. Color reconnaissance studies. National Bureau of Standards. Report 5184, 39 pp. U. S. Dept. of Commerce.

Keegan, H. J., J. C. Schleter and W. A. Hall. 1956. Spectrophotometric and colorimetric record of some leaves of trees, vegetation, and soil. National Bureau of Standards Report, No. 4528, U. S. Dept. of Commerce.

Keegan, H. J., J. C. Schleter and W. A. Hall. 1956. Spectrophotometric and colorimetric record of diseased and rust resisting cereal crops. National Bureau of Standards Report, No. 4591, U. S. Dept. of Commerce.

Kern, C. D. 1963. Desert soil temperatures and infrared radiation received by TIROS III: Atmos. Sci. 20:175-176, "Reply" - W. Nordberg and W. R. Bandeen, pp. 176-178.

Kimball, H. H. and I. F. Hand. 1930. Reflectivity of different kinds of surfaces. U. S. Monthly Weather Review 58:280-282.

Klesnin, A. F. and I. A. Sulgin. 1959. Ob opticeskih svojstvah listev rastenij (The optical properties of plant leaves.) Dokl. Akad. Nauk. SSR. 125: No. 5, pp. 1158-1161.

Knipling, E. B. 1967. Physical and physiological basis for differences in reflectance of healthy and diseases plants. Proc. Workshop on Infrared Color Photography in Plant Sciences. Winterhaven, Fla. 24 p.

Kohn, F. 1951. The use of aerial photographs in the geographic analysis of rural settlements. Photogr. Engr. 17:759-771.

Komarov, V. B., B. V. Shilin, M. M. Miroshnikov and Ju. A. Feoktistov. 1968. The methods of applications of infrared aerial photography when studying the volcanoes and thermal activities of Kamchatka Peninsula. Proc. 5th Symp., Remote Sensing of Environment, 821-822. Univ. Michigan, Ann Arbor.

Konaski, C. F. 1964. Water and cloud temperatures measured from U-2 aircraft. Amer. Meteor. Soc., Bull. 45:581-586.

Kondratiev, K. Y., Z. F. Mironova and A. N. Otta. 1964. Spectral albedo of natural surfaces. Pageoph 59:207-216.

Konrad, T. G. 1968. Radar as a tool in meteorology, entomology and ornithology. Proc. 5th Symp., Remote Sensing of Environment, 655-666. Univ. Michigan, Ann Arbor.

Kophlov, V. M. 1964. Calculation of the contrast of the optical image of features in an aerial camera. Am. Geophysical Union, No. 6, 355-358.

Krinov, E. L. 1947. Spectral reflectance properties of natural formations. Academy of Sciences, U.S.S.R., Translated by G. Belkov, Natural Research Council of Canada, 1953. Tech Transl. 439, p. 268.

Kumpf, H. E. and H. A. Randall. 1961. Charting the marine environment of St. John. U. S. Virgin Islands. Bull. Marine Sci. 11:543-551.

Lack, D. L. 1959. Watching migration by radar. British Birds 52:258-267.

Lack, D. L. Migration across the North Sea studied by radar.
1959. Part 1. Survey through the year. Ibis 101:209-234.
1960. Part 2. The spring departure. Ibis 102:26-57.
1962. Part 3. Movements in June and July. Ibis 104:74-85.
1963. Part 4. Autumn. Ibis 105:1-54.

Lack, D. L. 1962. Radar evidence on migratory orientation British Birds 55:139-158.

Langley, P. G. 1959. Aerial photography as an aid in insect control in western pine and mixed conifer forests. J. Forestry 57:169-172.

Latham, P. 1963. Methodology for an instrumented geographic analysis. Annuals Assoc. Amer. Geographers 53:194-209.

Lattman, L. H. 1963. Geologic interpretation of airborne infrared imagery. Photogr. Engr. 29:83-87.

Lawrence, P. R. 1957. Testing the efficiency of photo-interpretation as an aid to forest inventory. 7th British Commonwealth For. Conf.

Leedy, D. L. 1948. Aerial photographs, their interpretation and suggested uses in wildlife management. J. Wildlife Management 12:191-210.

Lee, W. T. 1922. The Face of the Earth as seen from the Air. Amer. Geographical Soc., N. Y., 110 pp.

Leestma, R. A. 1966. Application of air and spaceborne sensor imagery for the study of natural resources. Proc. 4th Symp., Remote Sensing of Environment, 111-113.

Legault, R. R. and F. C. Polcyn. 1965. Investigations of multispectral image interpretation. Proc. 3rd Symp., Remote Sensing of Environment, 813-821. Univ. Michigan, Ann Arbor.

Leighty, R. D. 1965. Terrain mapping from aerial photography for purposes of vehicle mobility. J. Terramechanics 2:55-67.

Leonardo, E. S. 1964. Capabilities and limitations of remote sensors. Photogr. Engr. 30:1005-1010.

Liang, T. 1964. Tropical soils: characteristics and airphoto interpretation, final report. Office of Aerospace Res. USAF, Bedford, Mass. AFCRL 64-937, 158 pp.

Lohman, S. W. and C. J. Robinove. 1964. Photographic description and appraisal of water resources. Photogrammetria 19:21-41.

Loomis, W. E. 1965. Absorption of radiant energy by leaves. Ecology 46:14-17.

Losee, S. T. B. 1942. Air photographs and forest sites. I. Mapping methods illustrated on an area of the Petawawa Forest Experiment station. Forestry Chronicle 18:125.

Losee, S. T. B. 1951. Photographic tone in forest interpretation. Photogr. Engr. 17:785-799.

Low, M. J. D. and F. K. Clancy. 1967. Remote sensing and characterization of stack gases by infrared spectroscopy, an approach using multiple-scan interferometry. Environment Sci. and Technology 1:73-74.

Lowe, D. S. 1968. Line scan devices and why to use them. Proc. 5th Symp., Remote Sensing of Environment. 77-102. Univ. Michigan, Ann Arbor.

Lowe, D. S. and F. C. Polcyn. 1965. Multispectral data collection program. Proc. 3rd Symp., Remote Sensing of Environment, 667-680. Univ. Michigan, Ann Arbor.

Lowe, D. S. and J. G. N. Braithwaite. 1966. A spectrum matching technique for enhancing image contrast. Applied Optics 5:893-898.

Lowman, D. 1965. Space photography - a review. Photogr. Engr. 31:76-86.

Lowman, P. D., Jr. 1967. Geologic applications of orbital photography. NASA Tech. Note D-4155, 37 p. Natl. Aeronautics Space Administration, Washington, D. C.

Lowman, P. D., Jr. 1966. The earth and orbit. Natl. Geogr. Mag. 645-671.

Lueder, D. R. 1959. Aerial Photographic Interpretation: Principles and applications. McGraw Hill, New York, 462 p.

Lukens, J. E. 1968. Color aerial photography for aquatic vegetation surveys. Proc. 5th Symp., Remote Sensing of Environment, 441-446. Univ. Michigan, Ann Arbor.

Lund, H. G., G. R. Fahnsetock and J. F. Wear. 1967. Aerial photo-interpretation of understories in two Oregon oak stands. U. S. For. Serv., Pac. Northwest For. Rang. Expt. Sta., Res. Note PNW-58.

Lutz, H. J. and A. P. Caporaso. 1958. Indicators of forest land classes in air-photo interpretation of the Alaska interior. U. S. For. Serv., Alaska For. Res. Center, Sta. Paper 10:1-31.

McClellan, W. D., J. P. Meiners, and D. G. Orr. 1963. Spectral reflectance studies on plants. Proc. 2nd Symp., Remote Sensing of Environment, 403-413. Univ. Michigan, Ann Arbor.

McLerran, J. H. and J. O. Morgan. 1965. Thermal mapping of Yellowstone National Park. Proc. 3rd Symp., Remote Sensing of Environment, 517-530. Univ. Michigan, Ann Arbor.

McConnell, P. and E. Garvin. 1956. Cover mapping a state from aerial photographs. Photogr. Engr. 22:702-707.

MacDonald, R. B. and D. Landgrebe. 1967. Remote sensing for agriculture and natural resources from space. Proc. Nat'l. Symp. Amer. Astronautical Soc.

Mardon, A. 1965. Application of microwave radiometers to oceanographic measurements. Proc. 3rd Symp., Remote Sensing of Environment, 763-779. Univ. Michigan, Ann Arbor.

Meier, H. K. 1962. On the use of infrared emulsions for photogrammetric purposes. (English translation by G. Richeer). Original paper in Bildmessung and Luftbildwesen, No. 1/1962:27-28.

Meier, M. F., R. H. Alexander and W. J. Campbell. 1966. Multispectral sensing tests at South Cascade Glacier Washington. Proc. 4th Symp., Remote Sensing of Environment, 145-159. Univ. Michigan, Ann Arbor.

Meyer, M. P. 1963. The quantitative method in forest aerial photo interpretation on research—approaches and limitations. Photogr. Engr. 29:937-941.

Meyer, M. P. and D. W. French. 1966. Forest disease spread. Photogr. Engr. 32:812-814.

Meyer, M. P. and D. W. French. 1967. Detection of diseased trees. Photogr. Engr. 33:1035-1040.

Miller, L. D. 1966. Location of anomalously hot earth with infrared imagery in Yellowstone National Park. Proc. 4th Symp., Remote Sensing of Environment, 751-770. Univ. Michigan, Ann Arbor.

Miller, R. G. 1957. The use of aerial photographs in forestry in British colonies. 7th British Commonwealth For. Conf.

Miller, R. G. 1960. The interpretation of tropical vegetation and crops on aerial photographs. Photogrammetria 16:230-240.

Miller, V. C. and C. F. Miller. 1961. Photogeology. McGraw-Hill Book Co., N. Y.

Moessner, K. E. 1957. Preliminary aerial volume tables for conifer stands in the Rocky Mountains. Intermtn. For. Rang. Expt. Sta., Res. Pap. 41.

Moessner, K. E. 1960. Training handbook. Basic techniques in forest photo interpretation. U. S. Forest Service. Intermtn. For. Range Expt. Sta. 73 p.

Moessner, K. E. 1961. Comparative usefulness of three parallax measurement instrument and interpretation of forest stands. Photogr. Engr. 27:705-709.

Moessner, K. E. and G. A. Choate. 1966. Terrain slope estimation. Photogr. Engr. 32:67-75.

Molineux, C. E. 1964. Aerial reconnaissance of surface features with the multiband spectral system. Proc. 3rd Symp., Remote Sensing of Environment, 399-421. Univ. Michigan, Ann Arbor.

Molineux, C. E. 1965. Multiband spectral system for reconnaissance. Photogr. Engr. 31:131-143.

Moore, R. K., and D. S. Simonett. 1967. Radar remote sensing in biology. BioScience 17:384-390.

Moore, R. T., M. C. Stark, and L. Cahn. 1964. Digitizing pictorial information with a precision optical scanner. Photogr. Engr. 30:923-931.

Morain, S. A. 1967. Field studies on vegetation at Horsefly Mountain, Oregon and its relation to radar imagery. Univ. Kansas, CRES Rept. 61-22, 19 p.

Morain, S. A. and D. S. Simonett. 1966. Vegetation analysis with radar imagery. Proc. 4th Symp., Remote Sensing of Environment, 605-622. Univ. Michigan, Ann Arbor.

Morain, S. A. and D. S. Simonett. 1967. K-band radar in vegetation mapping. Photogr. Engr. 33:730-740.

Morrison, A. and J. B. Bird. 1965. Photography of the earth from space and its non-meteorological applications. Proc. 3rd Symp., Remote Sensing of Environment, 357-363. Univ. Michigan, Ann Arbor.

Moss, R. A. and W. E. Loomis. 1952. Absorption spectra of leaves: I. The visible spectrum. Plant Physiology 27:370-391.

Moxham, R. M. and A. Alcaraz. 1966. Infrared surveys at Taal Volcano, Phillipines. Proc. 4th Symp., Remote Sensing of Environment, 827-843. Univ. Michigan, Ann Arbor.

Moxham, R. M. and others. 1965. Thermal features at Mt. Rainier, Wash. as revealed by infrared surveys. Geol. Survey Research 1965: U. S. Geol. Survey Prof. Pap. 525D: D93-D100.

Moyer, Ralph H. 1949. Use of aerial photographs in connection with farm programs administered by the Productive and Marketing Administration, USDA. Photogr. Engr. 15:536-540.

Muniz, S. 1964. Photo interpretation in the highway materials program of the U. S. Forest Service. Photogr. Engr. 30:966-970.

Meyers, V. I., L. R. Ussery, and W. J. Rippert. 1963. Photogrammetry for detailed detection of drainage and salinity problems. Trans. Amer. Soc. Agric. Engr. 6:332-334.

Myers, V. I., D. L. Carter and W. J. Ripperton. 1966. Photogrammetry and temperature sensing for estimating soil salinity. Internatl. Comm. Irrig. Drainage, 4th Congr. New Delhi, Question 19:39-49.

Myers, V. I., C. L. Wiegand, M. D. Heilman and J. R. Thomas. 1966. Remote sensing in soil and water conservation research. Proc. 4th Symp., Remote Sensing of Environment, 801-814. Univ. Michigan, Ann Arbor.

National Academy of Sciences. 1968. Space applications summer study 1967. interim report. Vol. 1, 65 p. Natl. Acad. Sci., Washington, D. C.

Neblette, C. B. 1927. Aerial photography for study of plant diseases. Photo Era Magazine 59:346.

Neblette, C. B. 1928. Airplane photography for plant disease surveys. Photo Era Magazine 60:175.

Nefedov, K. Ye. 1963. Hydrogeological mapping on the basis of aerial photographs. Doklady Akademii Nauk SSSR. 148:676-678.

Nelson, H. J. 1959. The spread of an artificial landscape over Southern California. Annals Assoc. Amer. Geogr. 49:80-99.

Nelson, E., E. Bradshaw and A. E. Wieslander. 1957. Photo interpretation of vegetation and soils in wild land areas of California. Proc. Soil Sci. Soc. Amer. 21:106-108.

Nobe, C. 1961. Use of airphoto interpretation in agricultural land economics research. Land Economics 37:321-326.

Nobe, Kenneth C. 1958. Results of a test of air photo interpretation as a tool in farm management and land economics research. Land Economics 34:271-275.

Nobe, K. C. and H. W. Dill. 1959. Evaluation of agricultural food damage by airphoto analysis of flood plain samples. Agric. Economics Research, Vol. 11, No. 4, U. S. Dept. of Agriculture, Wash., D. C.

Norman, G. G. and N. L. Fritz. 1965. Infrared photography as an indicator of disease and decline in citrus trees. Proc. Fla. State Horticult. Soc. 78:59-63.

Obatan, F. 1941. Sur la reflexion du proche infrarouge par les surfaces vegetales. Compt. Rend. 212:621-623.

O'Neill, H. T. 1953. Keys for interpreting vegetation from air photographs. Photogr. Engr. 19:422-424.

Olson, C. E., Jr. 1960. Elements of photographic interpretation common to several sensors. Photogr. Engr. 26:651-656.

Olson, C. E., Jr. 1963. The energy flow profile in remote sensing. Proc. 2nd Symp., Remote Sensing of Environment, 187-199. Univ. Michigan, Ann Arbor.

Olson, C. E. Jr. 1963. Infrared sensors and their potential applications in forestry. Michigan Academy of Science, Arts, and Letters, Papers. 50:39-47.

Olson, C. E., Jr. 1963. Photographic interpretation in the earth sciences. Photograph Engr. 29:968-978.

Olson, C. E., Jr. 1963. Seasonal trends in light reflectance from tree foliage. Archives Internationales de Photogrammetric (Transactions of the Symposium on Photo-Interpretation, Delft, 1962). 14:226-232.

Olson, C. E., Jr. 1964. Spectral reflectance measurements compared with panchromatic and infrared aerial photographs. Univ. Michigan (IST) Rept. No. 4864-8-T, 21 pp.

Olson, C. E., Jr. 1967. Accuracy of land-use interpretation from infrared imagery in the 4.5 to 5.5 micron band. Annuls Assoc. Am. Geogr. 57:382-388.

Olson, C. E., Jr. 1967. Optical remote sensing of the moisture content of fine forest fields. Univ. Michigan. IST Rept. 8036 1-F, 21 pp.

Olson, C. E., Jr., R. E. Good, C. A. Budelsky, R. L. Liston, and D. D. Munter. 1964. An analysis of light reflectance from tree foliage made during 1960 and 1961. University of Illinois Agr. Expt. Sta., Rept. ONR Proj. NR-387-025, Cont. 1834, 218 pp.

Olson, D. L. and J. T. Cantrell. 1965. Comparison of airborne conventional photography and scanned ultra-violet imagery. Photogr. Engr. 31:507 (abs).

Olson, D. P. 1964. The use of aerial photographs in studies of marsh vegetation, Maine Agr. Expt. Sta. Bul. 13, p. 62.

O'Neill, H., O. Schulte, A. Barwick, and T. Hart. 1950. An experimental
    system of keys for the interpretation of vegetation on aerial photog-
    raphy: Chesapeake Bay Area. Tech. Rept. No. 1. on Research Contract
    N6-ONR-25505, Catholic University of America.
Ory, T. R. 1965. Line-scanning reconnaissance systems in land utilization
    and terrain studies. Proc. 3rd Symp., Remote Sensing of Environment,
    393-398. Univ. Michigan, Ann Arbor.
Paijmans, K. 1966. Typing of tropical vegetation by aerial photographs
    and field sampling in northern Papua. Photogrammetria 21:1-25.
Parker, D. C. 1962. Some basic considerations related to the problem
    of remote sensing. Proc. 1st Symp., Remote Sensing of Environment,
    7-23. Univ. Michigan, Ann Arbor.
Parker, D. C. and M. F. Wolff. 1965. Remote sensing. International
    Sci. and Technol. 43:20-31, 73.
Plummer, G. L. 1968. Color infrared photography, land use patterns
    and plant sciences. Ga. Acad. Sci. 26:23-32.
Peake, W. H. 1959. Interaction of electromagnetic waves with some
    natural surfaces. Inst. Radio Eng. Trans. on Antennas and Propagation
    (Spec. Suppl.) Vol. AP-7, pp. S324-S329.
Pearman, G. I. 1966. The reflection of visible radiation from leaves
    of some western Australian species. Aust. J. Biol. Sci. 19:97-103.
Peterson, J. T. 1965. On the distribution of lake temperatures in central
    Canada as observed from the air. Dept. Meteorology, Univ. Wis.,
    Tech. Rept. 22, 35 pp.
Pierson, W. J. and others. 1965. Some applications of radar return data
    to the study of terrestrial and oceanic phenomena. Amer. Astronaut.
    Soc., 3rd Goddard Memorial Symp., Washington, D. C., March 1965,
    AAS No. 65-54.
Poley, J. Ph. 1965. Contrast enhancement in photogeology by selective
    filtering. Photogr. Engr. 31:368-375.
Pope, R. B. 1957. The role of aerial photography in the current balsam
    woolly aphid outbreak. Forestry Chronicle 33:263:264.
Prat, S. 1936. Botanical photography with infrared light. Biological
    Photographic Assoc. 4:191-201.
Rabideau, G. S., C. S. French, and A. S. Holt. 1946. The absorption and
    reflection spectra of leaves, chloroplast suspensions, and chloroplast
    fragments as measured in an Ulbricht sphere. Amer. J. Botany 33:769-
    777.
Radforth, N. W. 1963. Airphoto interpretation of organic terrain for
    engineering purposes. Proc. Symp. Photo Interpretation, Delft, 1962,
    507-513.
Raspolozhenskiy, N. A. 1964. An airborne spectrometer for study of the
    spectral brightness of landscape features. Am. Geophysical Union,
    Trans. No. 6, 358-360.
Raup, H. M. and C. S. Denny. 1950. Photo interpretation of the terrain
    along the southern part of the Alaska highway. U. S. Geol. Surv. Bul.
    963-D:95-135.
Remple, R. C. and A. K. Parker. 1965. An information note on an air-
    borne laser terrain profiler and micro-relief studies. Proc. 3rd Symp.,
    Remote Sensing of Environment, 321-337. Univ. Michigan, Ann Arbor.
Richter, G. 1966. Dictionary of Optics, Photography and Photogrammetry.
    Elsevier Publ. Co., New York, N. Y.

Risley, E. 1968. Remote sensing activities of committees of the National Research Council. Proc. 5th Symp., Remote Sensing of Environment. 941-944. Univ. Michigan, Ann Arbor.

Robertson, G. W. 1966. The light composition of solar and sky spectra available to plants. Ecology 47:640-644.

Robinove, C. J. 1965. Remote sensor applications in hydrology. Proc. 4th Symp., Remote Sensing of Environment, 25-32. Univ. Michigan, Ann Arbor.

Robinove, C. J. 1967. Remote-sensing potential in basic data acquisition. Water Resources Bul. 3:32-46.

Robinove, C. J. 1968. The status of remote sensing in hydrology. Proc. 5th Symp., Remote Sensing of Environment, 827-832. Univ. Michigan, Ann Arbor.

Rosenfeld, Aziel. 1962. Automatic recognition of basic terrain types from aerial photographs. Photogr. Engr. 28:115-132.

Rosenfeld, Azriel. 1962. An approach to automatic photographic interpretation. Photogr. Engr. 28:660-665.

Rosenfeld, A. 1965. Automatic imagery interpretation. Photogr. Engr. 31:240-242.

Ryker, H. C. 1933. Aerial Photography. Method of determining timber species. Timberman, 34(5): 11-17.

Sayn Wittgenstein, L. 1960. Recognition of tree species on air photographs by crown characteristics. For. Res. Branch, Canada Forestry Tech. Note No. 95.

Sayn Wittgenstein, L. 1961. Phenologic aids to species identification on air photographs. Forest Research Branch (Canada) Tech. Note No. 104, 26 pp.

Schneider, W. J. 1966. Water resources in the Everglades. Photogr. Engr. 32:958-965.

Schneider, W. J. 1968. Color photographs for water resources studies. Photogr. Engr. 34:57-162.

Schultz, O. W. 1951. The use of panchromatic, infrared and color photography in the study of plant distribution. Photogr. Engr., 17:688-714.

Schwarz, D. E. and F. Caspall. 1968. The use of radar in the discrimination and identification of agricultural land use. Proc. 5th Symp., Remote Sensing of Environment, 233-248. Univ. Michigan, Ann Arbor.

Sebestyen, S. 1964. Machine aided reconnaissance photointerpretation. 8th SPIE Technical Symp., August, 1963, Los Angeles, California. S. P. I. E. Journal, Vol. 2.

Sharp, J. V., R. L. Christensen, W. L. Gilman, and F. D. Schulman. 1965. Automatic imagery interpretation. Photogr. Engr. 31:223-239.

Shay, J. R. 1966. Some needs for expanding agricultural remote sensing research. Proc. 4th Symp., Remote Sensing of Environment, 33-36. Univ. Michigan, Ann Arbor.

Shay, J. R. 1967. Remote sensing for agricultural purposes. BioScience. 17:450-451.

Shul'gin, I. A., A. F. Kleshnin, and M. V. Lomonosov. 1959. Correlation between optical properties of plant leaves and their chlorophyll content. Moscow State University, Inst. Plant Physiology, Acad. Sciences USSR.

Shul'gin, I. A., A. F. Kleshins, M. I., Verbolora, and V. Z. Podol'nyi. 1960. An investigation of the optical properties of leaves of woody plants using SF-4 spectrophotometer. Plant Physiology (USSR) 7:247-252.

Shul'gin, I. A. and V. S. Khazanov. 1960. On the reflection of light as related to leaf structure. Doklady Botanical Science (USSR).

Simonet, M. and J. Van Roost. 1957. La prise de vue aerienne en infrarouge au Congo Belge. Institute Geographigue du Congo Belge.

Simonett, D. S. 1966. Present and future needs of remote sensing in geography. Proc. 4th Symp., Remote Sensing of Environment, 37-48. Univ. Michigan, Ann Arbor.

Simpson, R. B. 1966. Radar, geographic tool. Annls. Assoc. Am. Geogr. 56:80-96.

Sisam, J. W. B. 1947. The use of aerial survey in forestry and agriculture. Imperial Agricultural Bureaux, Aberystwyth, Wales. Joint Publ. 9, 59-67.

Singer, S. F. 1962. Forest fire detection from satellites. J. Forestry 60:860-862.

Smith, J. H. G. 1957. Forest history from aerial photographs. For. Chronicle 33:390-392.

Smith, J. H. G. 1960. Intensive assessment of factors influencing photocrusing shows that local expressions of photo volume are best. Photogr. Engr. 26:463-469.

Smith, J. H. G. and D. Bajzek. 1961. Photo interpretation provides a good estimate of site index of fir, hemlock, and cedar. J. Forestry 59:261-263.

Smith, J. H. G. 1965. Biological principles to guide estimation of stand volumes. Photogr. Engr. 31:87-90.

Smith, J. T. 1963. Color—a new dimension in photogrammetry. Photogr. Engr. 29:1-15.

Sonu, C. J. 1964. Study of shore processes with aid of aerial photogrammetry. Photogr. Engr. 30:932-941.

Sorem, A. L. 1967. Principles of aerial color photography. Photogr. Engr. 33:1008-1019.

Soules, S. D. 1963. Spectral reflectance photography of the earth from Mercury spacecraft MA-8. U. S. Weather Bureau, Meteorological Satellite Lab. Report No. 22.

Southern, H. N. and W. A. S. Lewis. 1938. Infrared photograph of arctic birth forest and fells. J. Ecology 26:328-331.

Spurr, S. H. and C. T. Brown, Jr. 1946. Specifications for aerial photographs used in forest management. Photogr. Engr. 12:151-161.

Spurr, S. H. 1948. Aerial photographs in forestry. Ronald Press Co. N. Y. 340 pp.

Steen, W. W. and J. C. Little. 1959. A new portable reflectance spectrophotometer for the selection of film and filters for aerial photography. Photogr. Engr. 25:615-618.

Steiner, D. 1960. Die Frage der gunstigsten Jahreszeit bei der Landnutzungs-interpretation auf panchromatischen Luftbildern (The question of the best season for the interpretation of land use on panchromatic air photographs). Communication to the IXth International Congress of Photogramm., Commission VII, London, England.

Steiner, D. 1965. Use of air photograph for interpreting and mapping rural land use in the United States. Photogrammetria 20:65-80.

Steiner, D. and H. Haefner, 1965. Tone distortion for automated interpretation. Photogr. Engr. 31:269-280.

Steiner, D. and T. Gutermann, 1966. Russian data on spectral reflectance of vegetation, soil, and rock types. Final Tech. Rept., Contract No. DA-91-EUC-3863/0I-652-0106, Department of Geography, University of Zurich, Switzerland, 232 pp.

Stingelin, R. W. 1968. An application of infrared remote sensing to ecological studies: Bear Meadows Bog, Pennsylvania. Proc. 5th Symp., Remote Sensing of Environment, 435-440. Univ. Michigan, Ann Arbor.

Stoeckler, E. G. 1949. Identification and evaluation of Alaskan vegetation from airphotos with reference to soil, moisture and permafrost conditions. U. S. Army Cold Regions Res. Engr. Lab., Tech. Rept. 21, 103 pp.

Stone, K. H. 1948. Aerial photographic interpretation of natural vegetation in the Anchorage area, Alaska. Geogr. Rev. 38:465-474.

Stone, K. H. Air photo interpretation procedures. Photogr. Engr. 22:123-132.

Stone, K. H. 1964. A guide to the interpretation and analysis of aerial photos. Annals. Assoc. Amer. Geogr. 54:318-328.

Strandberg, H. 1963. Analysis of thermal pollution from the air. Photogr. Engr. 29:656-671.

Strandberg, C. H. 1965. Aerial photographic interpretation techniques for water quality analyses. Photogr. Engr. 31:506 (abs.).

Strandberg, C. H. 1966. Water quality analysis. Photogr. Engr. 32:234-248.

Suits, G. H. 1960. The nature of infrared radiation and ways to photograph it. Photogr. Engr. 26:763-772.

Sweet, M. H. 1952. Color film analysis from integral density data. Optical Soc. Am. 42:232-237.

Tageeus, S. V. and A. B. Brandt. 1960. Changes in optical properties of leaves in the course of the growing season. Doklady Botanical Science (USSR).

Tarkington, R. G. and A. L. Sorem. 1963. Color and false-color films for aerial photography. Photogr. Engr. 29:88-95.

Tarkington, R. G. 1968. Photography: An electronic-ionic medium for remote sensing applications. Proc. 5th Symp., Remote Sensing of Environment, 3-8. Univ. Michigan, Ann Arbor.

Taubenhaus, J. J., N. Ezekiel and C. B. Neblette, 1929. Airplane photography in the study of cotton root rot. Phytopathology 19:1025-1029.

Texas Instruments, Inc. 1961. Infrared imagery for terrain analysis. Science Services Div., 100 Exchange Park, North Dallas, Texas.

Thomas, J. R., V. I. Meyers, M. D. Heilman, and C. L. Wiegand. 1966. Factors affecting light reflectance of cotton. Proc. 4th Symp., Remote Sensing of Environment, 305-312. Univ. Michigan, Ann Arbor.

Tomlinson, R. F. and W. G. E. Brown. 1962. The use of vegetation analysis in the photo interpretation of surface material. Photogr. Engr. 28:584-592.

Truesdale, P. E. 1959. Study of vegetation and terrain conditions from aerial photography. U. S. Naval Photographic Interpretation Center, Washington, D. C.

Tupper, J. L. 1956. The influence of atmospheric haze on the quality of aerial photographs. Photogr. Engr. 22:907-911.

Udall, S. L. 1965. Resource understanding - a challenge to aerial methods. Photogr. Engr. 31:63-75.

University of Michigan, Inst. Science and Technology. 1966. Peaceful uses of earth-observation spacecraft, Vol. 1, Introduction and summary, Vol. II Survey of applications and benefits, Vol. III Sensor requirements and experiments, Rept. No. 7219-1-F, NASA contract NASw-1084.

U. S. Army. 1962. Airphoto pattern reconnaissance of northwestern Canada. Vol. 1 and 2. U. S. Army Arctic Construction and Frost Effects Lab, Tech Rept. 41, 115 p. + 180 figs.

U. S. Naval Photographic Interpretation Center. 1950. Guide to Pacific landforms and vegetation, for use in photographic interpretation. U. S. Navy, Washington, D. C.

Van Lopik, J. R., A. E. Pressman and R. L. Ludlum. 1968. Mapping pollution with infrared. Photogr. Engr. 34:561-564.

Van Miegroet, M. 1965. Transmission and reflection of light by the leaves of some broadleaved species. Schweiz, Z. Forstwiss 116:556-589.

Vivian, W. E. 1963. Application of passive microwave techniques in terrain analysis. Proc. 2nd Symp., Remote Sensing of Environment, 119-125. Univ. Michigan, Ann Arbor.

Wagner, R. 1963. Using airphotos to measure changes in land use around highway interchanges. Photogr. Engr. 29:645-649.

Waldron, V. G., E. Marrow and D. C. Yates. 1960. Bibliography of reconnaissance interpretation, Volume 2. Science and Tech. Div., Library of Congress, Contract AF 30(602) 1835, Proj. 6244, RACD TR 60-150B. 90 pp.

Waldron, V. G., W. Colson, and D. C. Yares. 1961. Bibliography of reconnaissance interpretation. Volume 3. Science and Tech. Div. Library of Congress, Washington, D. C. 115 p. (Contract AF 30 (602)2251, Proj. 6244, Continuation of CONTRACT AF 30(602) 1835).

Walker, J. E. 1961. Progress in spectral reflectance film-filter research applicable to engineering and geologic studies. Photogr. Engr. 27:445-450.

Walker, T. J. 1965. Detection of marine organisms by an infrared mapper. Woods Hole Oceanographic Inst., Oceanography from Space. Ref. 65-10.

Waters, W. E. and R. C. Heller, and J. L. Bean. 1958. Aerial appraisal of damage by the spruce budworm. J. Forestry 56:269-276.

Wear, J. F., R. B. Pope and P. W. Orr. 1966. Aerial photographic techniques for estimating damage by insects in western forests. U. S. For. Serv., Pac. Northwest For. Rang. Expt. Sta.

Weber, F. P. 1966. Multi-spectral imagery for species identification. Pacific Southwest Forest and Range Experiment Station. Berkeley, California, For. Remote Sensing Lab, Final Rept. NASA Cont. R-09-038-002, 37 pp.

Weber, F. P. 1966. Exploration of changes in emitted and reflected radiation properties for early remote detection of tree vigor decline. Mich. Acad. of Sci., Arts, & Letters. 17th Annual Report.

West, D. F. 1959. Aerial surveys for better crops. Crops and Soils 11:17-18.

Wieslander, A. D. and R. C. Wilson. 1942. Classifying forest and other vegetation from air photographs. Photogr. Engr. 8:203-215.

Whittlesey, H. 1966. Bipod camera support for close-up photogrammetry. Photogr. Engr. 32:1005-1010.

Wickens, G. E. 1966. The practical application of aerial photography for ecological surveys in the savannah regions of Africa. Photogrammetria 21:33-41.

Widger, W. K. 1966. Orbits, attitudes, viewing geometry, coverage, and resolution pertinent to satellite observations of the earth and its atmosphere. Proc. 4th Symp., Remote Sensing of Environment, 489-538. Univ. Michigan, Ann Arbor.

Wiegand, D. L., M. D. Heilman and A. H. Gerbermann. 1968. Detailed plant and soil thermal regime in agronomy. Proc. 5th Symp., Remote Sensing of Environment, 325-342. Univ. Michigan, Ann Arbor.

Wieslander, A. E. and R. Storie. 1953. Vegetational approach to soil surveys in wild land areas. Proc. Soil Sci. Soc. Amer. 17:

Wilson, A. 1966. The remote surveillance of forest fires. Applied Optics 5:899-904.

Wilson, R. A. 1968. Fire detection feasibility tests and system development. Proc. 5th Symp., Remote Sensing of Environment, 465-478. Univ. Michigan, Ann Arbor.

Wilson, R. C. 1948. Photo interpretation aids timber surveys. J. Forestry 46:41-44.

Wilson, R. C. 1949. The relief displacement factor in forest area estimates by dot templates on aerial photographs. Photogr. Engr. 15:225-236.

Wilson, R. C. 1962. Surveys particularly applicable to extensive forest areas. Proc. 5th World For. Congr. Nat'l. Publ. Co. Washington, D. C.

Wilson, R. C. 1966. Forestry applications of remote sensing. Proc. 4th Symp., Remote Sensing of Environment, 63-70. Univ. Michigan, Ann Arbor.

Wilson, R. C. 1967. Space photography for forestry. Photogr. Engr. 33:483-490.

Wimbush, D. J., M. D. Barrow and A. B. Costin. 1967. Color stereophotography for the measurement of vegetation. Ecology. 48:150-153.

Woods Hole Oceanographic Institution. 1965. Oceanography from space—proceedings of conference on the feasibility of conducting oceanographic explorations from aircraft, manned orbital and lunar laboratories—collection of papers. 24-28 August 1964.

Worley, D. P. and M. E. Dale. 1960. Recording tree defects in stereo. Central States For. Expt. Sta., Tech. Pap. 173, 11 pp.

Wright, S. 1939. The aerial photographic and photogrammetric activities of the Federal Government. Photogr. Engr. 5:168-176.

Yost, E. F. and S. Wenderoth. 1967. Multispectral color aerial photography. Photogr. Engr. 33:1020-1033.

Yost, E. F. and S. Wenderoth. 1968. Coastal water penetration using multispectral photographic techniques. Proc. 5th Symp., Remote Sensing of Environment, 571-586. Univ. Michigan, Ann Arbor.

Zsilinszky, V. G. 1964. The practice of photointerpretation for a forest inventory. Photogrammetria 19:42-58.

Zsilinszky, V. G. 1966. Photographic interpretation of tree species in Ontario. Ontario Dept. Lands and Waters, 86 pp.